Ham Radio

Tips and Techniques for the Perfect Signal

(The Ultimate Ham Radio Complete Guide the Easiest Way)

David Shanklin

Published By **Phil Dawson**

David Shanklin

Ham Radio: Tips and Techniques for the Perfect Signal (The Ultimate Ham Radio Complete Guide the Easiest Way)

ISBN 978-1-9994868-5-3

No part of this guidebook shall be reproduced in any form without permission in writing from the publisher except in the case of brief quotations embodied in critical articles or reviews.

Legal & Disclaimer

Table Of Contents

Chapter 1: Let's Start With Ham Radio
The History of Ham Radio

It seems that history cannot keep the name of the radio amateur who was first discovered. The experiments conducted "for themselves" and then reported of wireless telegraphy weren't kept in the records. In general the only family and friends of the home scientist were aware of them. One of the pioneers was Slovak religious priest Josef Murgash. He emigrated in 1896 in the USA and was intrigued by the radio's nascent technology, which was which was equipped with a laboratory at home in which he conducted the experiments of radio and telegraphy. His inquisitive brain suggested that he cut down on the transmission time of messages telegraphed in Morse code. The idea was to transmit "dots" as well as "dashes" with parcels with the same brief duration, but at different frequency. Murgash made two applications for this technique - both in 1903 as well as 1904 and in 1904 he received two patents.

In the present time The fading notion of "radio amateur" as well as "HF radio sportman" about 15 years ago was fascinating. People with a soldering iron at their disposal, they took apart circuits, and shared ideas, as well as new designs for transceivers. They displayed remarkable ingenuity and creativity as they sought out the required radio parts in the context that were characterized by a general shortage

How Far Can Amateur Radio Be Established?

The distance of the transmitter's signal is not determined just by the power of it but by the conditions that govern the transmission by radio waves (passage). In the right conditions for propagation by the HF transmitter and even with very little power (units as well as fractions of Watt) is able to be heard over the distance of almost all arbitrarily long distances. In particular, in the 28 or 21 MHz bands, communications between Antarctica or Australia by using a transmitter that has a power between 10 and 20 W as well as an ordinary antenna is uncommon or even communications between American and Japanese radio amateurs is usually an everyday routine particularly during periods of high solar activity. In VHF Long-range propagation radio waves is extremely uncommon and usually restricted to Europe. However, with a lot of experience and high-quality equipment Some radio amateurs routinely create long-range communication using VHF as well as the HF (such as, for instance, reflection of radio signals coming that come from to the Moon). Research into the pattern that radio waves is among the most fascinating and important scientific tasks that radio amateurs engage in.

How Expensive Is Amateur Radio?

As high as you're able to spend. A high-quality home-built device can be built by yourself using low-cost radio components in the end, most of the price will be made up of your labor put into the process.

Even if you do have lots of cash and want to spend it on the purchase of various antennas, equipment as well as industrial-grade accessories can be done easily. The prices for shortwave amateur transceivers (transceivers) produced by industrial companies (mostly Japanese) range from 600 to 9000 dollars, as well as a variety of HF antennas, ranging between 200 and 3000 USD. An inexpensive portable 2m VHF transceiver is priced at approximately $150. A used model can be less than half.

Is There Any Practical Use For Amateur Radio?

The practical applications in Ham Radio can be described by the following criteria:

1. A brief introduction to the technical understanding of students and, more

generally, individuals from diverse occupations.

2. Acquiring practical experience for handling electronic devices that are complex.

3. Assistance with emergency communication during times of natural catastrophes, as well as artificial catastrophes.

4. Communicating in circumstances that are not typical (expeditions and rescue operations).

5. Mass research.

6. The general perspective of individuals with the various degrees of education.

7. Strengthening mutual understanding between people.

Can Radio Emissions From An Amateur Transmitter Be Harmful To Health?

Legal amateur radio transmitters haven't been reported. The typical mobile phone can be possibly more risky and so are the high voltage power lines. In the heads of uninformed laypeople, huge antennas are frequently viewed as the risk of health problems however, these beliefs have none of the scientific evidence. An antenna alone is not able to affect the health of a person (after all, they are similar metal structures that we all have). The answer is dependent on the type of radio waves are being emitted and at what intensity and the frequency at which they are utilized.

Furthermore, when radio signals occur through the antenna, the greater the antenna's location more likely is to have an impact to the surrounding environment. Amateur radio stations generally produces no emissions as it is situated at reception. For the brief time where there is transmission, the parameters for radio emission do not exceed the standards set by the sanitary standards. But, regardless it is imperative that safety standards be observed to the letter.

The Hardware Of A Ham Radio Station

The Ham HF radio station's equipment may comprise separate receiver and transmitter unit, and also the combination device is known as transceiver. Additionally, their combinations can be found in a variety of situations, especially during competitions. In general, the transceiver has been designed to work with every amateur radio communication band. It is able to cover the frequency that ranges from 0.1 to 30 MHz however, for novice radio amateurs using the 1-2 band device, it can be sufficient for understanding the fundamentals. However, it's a good idea to own a multi-band receiver. The frequency of both the receiver and the transmitter of the transceiver share the same frequency making it easier to operate in the air.

Amateur equipment is able to be constructed in a self-contained manner, however this will require an appropriate qualification for an amateur radio operator, prior previous experience building equipment in this category, and the ability to measure. Like any other enterprise, the gear that is used in the workplace is vital.

To this end, an person who is watching only needs an antenna, however sporting results are contingent on the receiver's quality, and foremost sensitiveness of the receiver's path of reception. The first receiver I used was the tube receiver of a radio station in the military; the device was heavy, highlighting the power of the troops with my armor. In the next step, utilizing the soldering iron, modern receivers were constructed using the magazine "Radio" along with "Radio Amateur,"" and immediately affected reception.

Equipment And Apparatus

To broadcast FM there is a need to choose the space and install with a studio. There aren't any specific requirements to arrange rooms. The most important thing is to provide adequate sound insulation in order so that you can avoid disturbances in the air. This will allow the radio station to be a tiny regional one to purchase an apartment in the upper floor of a multi-story structure with the area of 50 square metres. M. or a space that is suitable.

More significant and costly is the equipment you choose. For the installation of antennas it is suggested to talk with the owner of the tower for a discussion on the location of your antenna since it is possible to install by purchasing an area of land, providing electricity, and then directly building and maintaining this tower could require millions of dollars expenditures. Transmitters do not have these costs like inside an radio station.

The complete list of essential equipment is comprised of the following items:

1. Console

2. Recorder

3. Cable-based feeder

4. Microphones, headphones

5. Equipment for the Auxiliary

6. Equipment for taking telephone calls and placing them over the air

7. PC Office equipment, PC, software.

It's not the best performance and quality however, it is the primary equipment must be used by the station. It is possible to vary the equipment based upon the qualifications of employees as well as the characteristics of the radio station. In any event these devices should be connected, mounted, and launched by professionals in order to perform seamlessly.

What Is Radio Technology?

Heinrich Hertz was the first to show how electricity as well as magnetic field could be used to transfer messages. At the time, 1886 the technology was not yet available in transistors or electron tubes. Transmission signals were produced by spark gaps. It was only later found out that the power of signals, also known as radio waves, can be extended by using longer wires. The spark gap dwindled and the antenna was created with the term Funk stuck. That's the reason that every technological establishment that utilizes electric and magnetic fields to transfer messages or data is classified by radio as a technology.

What Is An Antenna?

An antenna is a piece equipment which can transmit and receive electromagnetic signals. Its main function is as a transducer made of metal between lines and space. This allows the transmitting or receiving of signals. While it is connected as a two-pole device, its fundamental structure comprises four poles, two of them aren't necessarily fixed in any way, and instead hang free throughout space.

How Are Radio Waves Formed?

Radio waves need a vibration generator for their creation and without this, they can't exist. Generators are typically an oscillator, which creates an essential or carrier wave which is a physical phenomenon. But in the realm of electronic technology, the word "waves" does not typically utilized to mean radio waves since they're frequency. An oscillator produces the AC voltage signal at particular frequency. once the frequency is at the level of a certain amount the electrical signals are able to propagate into space. The frequency is determined as Hertz (Hz) which is which measures frequencies.

Basic Components Of A Radiofrequency System

The basic radiofrequency system consists of two fundamental elements: the radio transmitter or transceiver, as well as the transmission line which includes cables, connectors, as well as an antenna. Transmitters come in all kinds and uses including data transmission as well as radio, voice television.

Every system has to meet specific requirements, like the impedance, the signal strength as well as frequency response, gain and attenuation levels. A transceiver or radio is is accountable for the generation or reception of the intended signal. It is still essential to transmit it via the transmission line, which consists of connectors, cables as well as an antenna.

The cables must be of the finest quality that is able and constructed to function with radio frequencies, based on the equipment that it is joined and the climate conditions on the area. There are various kinds of cables such as coaxial LMR as well as Heliax and Heliax, which are the most commonly used. Each is different in losses or attenuations at specific frequency and lengths, which is expressed in dB/m. An antenna transmits radio waves or RF energy through space. The antenna is chosen based on the customer's requirements, frequency of operation an impedance value, the dimension, pattern of radiation, as well as the cost.

The omnidirectional and the directional grids and parabolic mast reflections. Billiard cues, as well as a range of material and gain; the second is measured in decibels (decibels) (decibels) or in decibels (isotropic equivalent decibels). Don't confuse dB with dBi as there's the difference between the two of 2.15 units. That is an antenna with a 4dB gain is the same that the 6.15 dBi one. A lot of companies describe the gain of their equipment as dBi in order to suggest that they gain more, however this isn't the case.

Connectors join cables, the radio and antenna. There are a variety of connectors including UHF or SMA, PL BNC, TNC etc. Each of them comes with variations with female, male and reverse. In the same manner we recommend to choose the highest quality possible since the wrong or unreliable connector could cause issues with our transmission lines.

How To Make An FM Receiver At Home?

The most fascinating subjects for fans of electronics is the transmission and receiving in radio signal. Assembly of the radio receiver can be relatively straightforward as long as you don't require high-sound quality. We propose that we build a sensitive radio receiver within the commercial range of modulated frequency. Let's do it!

Functioning

In this figure in the figure, you will observe the simplified scheme of the receiver. It can be split in two major blocks in the following manner:

1. Radio frequency section

2. Section Low-frequency

Section Of Radio Frequency

The Q1 transistor and its connected components comprise Radio frequency sections. The transistor functions to function as an oscillator and mixer. The capacitor C2 generates positive feedback to the collector and emitter. This causes an oscillation in the transistor, with an interval which is determined by the tune circuit that is formed by the C1 and L1. This frequency can be altered via CV and lies located in the FM range. Capacitors C3 as well as VR11 create the stop and oscillation of the transistor at a very low frequency (not loud) that is higher than 20Khz. The continuous stop and start creates an environment of extremely sensitive that the transistor has to frequencies similar to the frequency in which it oscillates.

Resistors R1 & R2 as well as VR1 make up the bias circuit of the transistor Q1. Capacitor C1 is the one responsible to decouple the power supply line that is positive. The radio signal that is captured in the antenna gets transmitted through C4's step capacitor to the tuned circuit created by CV and L1. This circuit will be mixed with the sound generated by the oscillator that is similar frequency to the recorded signal. Thus, the low-frequency (sound) altered in the radio signal creates distinct signals.

In order to avoid overloading on the circuit tuned (which could result in null selection) The low-frequency signal originates via the Q1 emitter. The 1mH L2 coil functions as a high-impedance choke in front of any radio frequency signals hindering its transfer from the emitter to ground. Audio signals are derived by a pi filter that is formed by capacitors C5,C8 as well as resistors R3.

Resistor R4 has the task in limiting the use of Q1 so that it is prevented from generating excessive energy that can cause interference with the receivers in close proximity.

Like you see, the process of tuning and modulating a radio frequency signal in order to get the sound using a basic oscillator stage is easy. A lot of readers will know that this particular type of receiver goes by its name "reaction receiver." It's been in use for a long time, even before the invention of radio.

Section Of Low Frequency

When the audio signal has been received, amplifying it so that you can listen to it with the earphones or speakers with low impedance is essential. One of the people responsible for amplifying this signal is integrated circuit LM386, which comes with an 0.5 W amplifier that is only 8 pins. The P1 potentiometer acts as the control for volume of our tiny receiver. Capacitors, whereas C11 and C10 transfer the audio signal, but they prevent the DC component from being blocked. To hear, it's essential to connect a speaker or headphones that are between 8 and 32 ohms, to the amplifier's output.

Assembly And Adjustment

In order to assemble the receiver, a fine tip soldering iron essential, along with 1mm of tin wire, and cutting pieces. The components are inexpensive and easy to locate. For a circuit board we suggest making it in fiberglass. But, it might be appropriate to utilize an island plate similar to that used in our lab for tests and adjusts of the circuit prototype. It's not recommended to build the circuit using prototype boards (Board) (Board) or "line" boards because each has a huge capacity for connecting adjacent lines which can result in a malfunctioning of the receiver. When you are assembling the circuit ensure that you leave enough space to accommodate the ground and power lines and then solder the circuit closest to the boards as you can (short wires).

The L1 coil is constructed by winding 4 turns of conductor wire around an 0.8mm center. An ordinary 9v battery provides power to the receiver. After the receiver is assembled and the circuit is complete, placing it together with the battery inside the form of a box made of metal that needs to be electrically connected to a ground power source (ground) is recommended. This creates a barrier which blocks tuning fluctuations and unwanted noise.

It is possible to make small holes inside the enclosure for tuning or volume adjustments. To make an antenna, you can make sufficient with a thread that is about 20cm. Adjustment: Adjusting the receiver is a breeze. Adjust the volume knob half, then turn it up until you can hear the usual rattle of an untuned receiver emanating from your speaker. Then, alter the condenser's CV until you are able to tune into a station. If there is no sound you can try a slight adjustment of VR11 or turn the turns of L1 further apart or closer together.

Enhancements: The circuit described in this article is suitable for reception of all other signals that falls within the spectrum of medium and short wave. Increase the values of L1 and CV to reach 10 turns for L1 and at least 500 pF in CV is required for receiving the signals in these bands. It is capable of detecting the signals regardless of their frequency of modulation (FM as well as AM).

Principle Of The Radio

Radio transmission is the process of making a radio signal with the right frequency and strength that is modulated by another frequency carrier and sending it out through an antenna. The signal that is received by the antenna is processed and then demodulated in order to get the signal that was originally sent, and possibly with some differences. Radio frequencies are classified into different frequencies based on frequency and wavelength, which includes medium, short, long high, ultrashort extremely high and hyper-high frequency.

Radio waves possess distinct properties and propagation laws dependent on the frequency of their propagation. The ionosphere absorbs large waves (DWs) while ground waves play a crucial role in propagation. The short waves (SWs) absorb during the day, and reflect during the evening by the ionosphere producing a radiation area within the transmitter. The propagation of VHF waves is along a straight line, and are able to bend over obstructions, whereas HHF waves are not bending extend beyond the sight line.

Chapter 2: Signals & Operations

How Is Data Transmitted By Radio?

Many people know that radio communications utilize electromagnetic waves for transmitting information to space, there's often some confusion regarding how these waves transfer data. Particularly, many people do not know how radio signals can be sent and received, especially regarding the transmission of vocal signals. Sounds travel slow through the air, and then quickly degrade, making the transmission of audio over distances of a long distance complicated. But, the process of converting sound waves into an electrical signal with a microphone can make transmitting the signal across several kilometers or more in a matter of minutes. If someone talks through a microphone, the power output of the device is altered to reflect the audio waves. However, how does the electrical signal transmitted across vast distances?

Through an antenna it's possible to discharge various electromagnetic waves into the air. The antenna's dimensions should be adequate to allow the effective transmission from radio frequencies. Particularly, the size of the antenna has to match the frequency of electromagnetic waves so that they can be radiated effectively. The frequency of sound waves can vary from 20 Hz up to 20.000 Hz. The frequencies ranging from 15 to 15,000 kilometers. Building an antenna of this magnitude is an arduous project. This is why sending audio signals directly to the air isn't a simple process. If radiation does occur in a certain frequency, the frequencies of audio signals that are broadcast by different stations are almost identical. Thus, the audio signals blend in the air and listeners are unable to select which one to listen to.

The male voice does not sound as appealing in comparison to the female voice while female voices are much generally more popular. This is the reason for an idea that when certain conditions are met high frequencies travel further than lower frequency. In order to illustrate this idea researchers have compared walking with driving in order to get there. Walking is more slow and takes an effort to walk in comparison to driving. The same is true for the transmission of low-frequency audio signals takes greater power, and it is also only limited in the distance. In order to solve this problem, different frequencies of high-frequency electromagnetic signals are produced for transmitting audio signals using antennas. The result is that multiple transmitters can utilize different frequencies of high-frequency electromagnetic waves, without affecting one another. In the end, small antennas can be developed to serve this function.

The validity of this idea was proven by practical experiments. The radio transmitter is a combination of many components in order to generate an electromagnetic wave that is which is then transmitted through an antenna. The transmitter, in particular, produces high-frequency electromagnetic waves. It transforms electricity and sound and manages low frequency, and then transmits the synthetic waves via the antenna. The high-frequency oscillator creates high-frequency waves, while embedding the audio signal into the high frequency vibrational wave. This process is termed modulation. After the modulation process the high-frequency vibrational signal is known as"an "already adjusted signal." The signal modulated transferred to an antenna using the transmission line. It allows it to radiate in a different site.

The transmitter is accountable of performing four essential tasks:

1. The conversion of audio into an electrical signal.

2. The production of high-frequency oscillating waves using an exact power.

3. Utilizing audio signals to control a specific characteristic of the high frequency oscillatory waves is called modulation.

4. Sending electromagnetic waves.

In order to perform the four duties that were mentioned earlier A transmitter has to comprise comprising several parts, such as a converter for transmitters or RF oscillator the modulator, RF amplifier as well as a transmitting antenna and power supply. When the carrier signal is transmitted The goal of a receiver is to record the electromagnetic signal transmitted in the air and return it back to the original frequency.

In order to receive electromagnetic signals to receive electromagnetic signals, an antenna called a receiver can be used. But, as many radio stations emit signals, an antenna can receive those radio signals you want as well as other signals with different frequency. Radio waves are broadcast by using different frequencies of carrier in order to identify the signals. The receiver can "select" the type of signal it wants to receive by comparing it to the appropriate carrier frequency.

Selective Circuit

A "selective circuit" in a receiver can't select a particular electromagnetic wave to be received from a specific station. It's also not possible to direct transfer the audio signal directly to an earpiece. The higher frequency "imposed" is to be eradicated as well as the original audio signal has to be returned. The procedure of detaching sound waves from electromagnetic waves is known as demodulation. The instrument that is used to accomplish this function is referred to as a demodulator. In order for the signal to be listened to, the audio signal generated by the demodulator has to be delivered through the headphones in order to allow the communication.

OOK (On-Off Keying)

"OOK" (On-Off Keying) is the most fundamental type of digital code that can be described as switching the transmitter off and on with a binary code. Because it is simple its widespread use for wireless remotes, radio buttons, as well as other affordable equipment. Most of the time, no encryption is used, and the sequence of bits and frequencies are wired and allow any user to send and receive signals. This means that it's not ideal to be used for securing expensive objects, like an expensive Lamborghini stored in garage. It is however enough for tasks that are less demanding, like managing a nightlight near the mattress. As an example, I've used a nightlight bought from a local minimart that uses the OOK principle over the last three years, with no false positives. It is an example of how to apply the "elusive Joe" principle in practice.

Amplitude modulation (AM)

AM modulation is anticipated to continue to be used for quite a long time, since it's utilized within broadcast stations as well as transmitters operating within the 118-137 the MHz band. One of the distinctive features of AM is the fact that the spectrum of AM is aligned with respect to its central frequency. This making it possible to determine if music or speech are transmitting in the form of a circle. AM was in fact among the first methods that was used to transmit and receive speech. The well-known "school" detector receiver circuit that was very simple and didn't require batteries to receive, made use of radio waves' energy to power headphones with high impedance. It is interesting to note that these receivers were not manufactured in mass quantities in the 1960s.

Modulation of a single sideband (USB, LSB, SSB)

Single sideband modulation, commonly called USB, LSB, or SSB is an original type of amplitude modulation. When AM signals are broadcast the spectrum is symmetrical to the central point. But, when using SSB modulation the spectrum is symmetrical, but just "half" part of the spectrum is transmitted giving a larger spectrum with the same transmitting energy.

FM stands for Frequency Modulation (FM)

Frequency modulation is the main principle for FM broadcasting. This sends out a layered signal which can include both stereo and mono channel, a pilot tone RDS as well as other. In order to distinguish the FM broadcasting in comparison to "regular" FM signals, the engineers often refer to it by the name of WFM (Wide FM). The entire spectrum of the radio station's signal is easily visible with the HDSDR application that shows the tone of the pilot at 19 kHz. It also shows RDS mono, stereo, and AM broadcasting channels. However the devices such as baby monitors and walkie talkies make use of "narrow" FM (NFM, Narrow FM) modulation where only audio is broadcast. Frequent modulation can also be utilized for digital signals which allow binary codes to be sent by switching between two frequency. With a low intensity, radio enthusiasts can communicate short messages over lengthy distances with FT8. For aircraft, an ACARS system can combine different modulation methods through the transmission of digital FM signals using AM transmitters, possibly due to the savings in cost from the use of already-existing equipment.

Phase Modulation (PSK)

Alongside the frequency modulation method, the phase modulation method is another option to alter a signal. You can ensure the reliability of communication across long distances through altering the phase of the signal, especially for satellite communications. In terms of radio protocol for amateurs, PSK31 was widely used for a short time. In PSK31 it is possible for a transceiver to be linked to a PC for exchange of information via text-based chat. The quantity of possible phases will vary based on frequency and the communication channel like 4, 18 and 16.

Modifying a signal's frequency and amplified amplitude at the same time is feasible which results in a more efficient transmission. This involves more complicated encoding and decoding. One example of this sign is QAM. The QAM signal type is easily visible by using a phase plane.

A Simple Shortwave (HF) Receiver For Beginners

We are shortwave radio amateurs with transmitters for shortwave radio and receivers available. Every radio amateur station has an individual call number, through which it's possible to identify the country in which or even the part of the country that station is situated. Radio amateurs using shortwave establish communication to one another and communicate with each other via a specific international "radio language" also known as, or in the case of radio"radio code," that is simple to recall.

By using codes tables, shortwaves are able to discuss topics that are connected with their equipment as well as their propagation. In addition, they can provide most relevant information to any shortwave on the audibility level from their broadcast station as well as the transmitting quality. Amateur radios typically have small power, less than 100 Watts. Amateurs make use of narrow portions of all shortwave frequencies. The narrow sections are known as "amateur bands" that are located within the bands of the 10-14 2040 as well as 130-meter band. In the example above the 40-meter amateur range extends from 41.6 to 42.8 meters, while the 20-meter band covers an area extends from 20.8 to 21.4 meters.

Utilizing a variety of waves Shortwaves are able to establish communication across short distances, and also far-reaching communications across thousands of miles. A receipt card that is specially designed is used to verify each radio amateur contact. A lot of radio amateurs own at their disposal hundreds of thousands of these receipt cards. Only an experienced radio enthusiast with a good understanding with radio circuits and can mount the radio equipment and set up radio communication, can create a radio shortwave station.

Building An Amateur Transmitter

For the construction of an amateur radio transmitter, an authorization by the Ministry of Communications is necessary that is granted in accordance with the relevant documentation. You can also get a shortwave license without the need for a transmitter by creating only one receiver. There are many these beginner shortwaves across the United States. Shortwave observers are also known as. Observers are not able to establish radio contacts for themselves, but they do are able to monitor radio communications in two directions between amateur shortwave stations in every amateur band. Being able to learn about an ultra-long distance station is as interesting as contacting it. It is a pleasure to receive distant stations is like radio broadcasting.

A shortwave observer who constructed a receiver for shortwave and is familiar with the rules of exchange between amateur radio stations can sign up his device at a radio club in the local area. The radio club will issue an individual certificate that identifies the specific shortwave receiver's "call symbol." In the event of a particular call sign, a observer of a shortwave can issue receipt cards to any amateur radio stations that he can hear in his receiver. Shortwaves, after receiving an observer's card and then sends the card as a response. In which he affirms the reception of the card from the observer along with a message regarding the frequency and quality from his radio station.

In the cards of observers among the observer cards, there is the card of one of the shortwaves from in the Far East or the Arctic as well as a card featuring radio amateurs of Prague Warsaw, and cards that feature shortwaves from France, Italy, India, Australia, the islands of Oceania and more. Short-wave radio amateurs operate mainly through radiotelegraphy. Therefore, in order to hear the activities of amateur radio stations, they need to know how to hear the alphabet of telegraphy, or at a minimum, with a lower rate of forty to fifty letters per minute. The study of the telegraph alphabet within circles or classes to shortwave radio users are the most effective.

It is possible to form a circle for studying what it takes to read the alphabet telegraphic through ear in school or at the House of Pioneers, or perhaps train with a buddy to receive by ear, and transmission via the key. If you are lucky it is possible to hear and receive independently, observing to the radiotelegraph stations in your radio, however it usually takes a lot of time and effort.

The detector phase of the receiver is constructed in a 3-point pattern that includes an anode which is grounded at a very high frequency. The circuit coil connects with one of the three triodes in the lamp using three points three points: a, b and the third one.

at point b, the coil connects via the grid R1C3 to the grid of the triode, the point A is directly connected to the cathode of the triode and point c to the ground where it's also linked via capacitance C4 as well as the anode from the triode.

A certain point at the axis of a coil of the circuit, this circuit produces oscillations i.e. the receiver becomes an engine. Stations must be heard close to the threshold of generation, the telegraph signal is received above the threshold for generation, at the point that the receiver just began to generate, as well as calls are exchanged before reaching the threshold of generation, indicating that the receiver is still creating. The more close to the threshold a station's signal is heard, the more it's heard, as well as the higher number of stations a receiver will be able to receive. This is why it's vital that the production of the receiver, also known as like they say feedback, be adjusted.

The feedback adjustments inside the receiver needs to be as easy as is possible. There are numerous ways to adjust the quantity of feedback. However, they can be all more or less complex and hard to implement. Our receiver employs an unique method to adjust the level of feedback. The feedback is controlled by factors! The resistance is between the first triode's cathode as well as the ground. The resistance R, which is attached to the component in the circuit for oscillation adds losses and increases the frequency of attenuation in the circuit.

If you adjust the amount of the resistance to be adjusted, the device will be set to an output that is in line with the threshold for generation.

The circuit is easy to build and doesn't need an additional variable capacitor to the receiver, as do other circuits to adjust feedback. It has been proven through tests that it provides a smooth transition to the threshold for generation and adjusts the circuit for receiver less than when you adjust the feedback of circuits using an capacitor. The voltage that is alternating in the audio frequency gets assigned to the load resistance of the anode R5, which is a part of the circuit for the anode of the left-hand triode. Through the capacitor Sv the voltage is fed to the grid in the second triode that acts as an amplifier for audio frequencies. The resistance R4 represents the loss of grid.

In order to maintain a constant negative bias to the grid of the second triode, resistance R5 is incorporated into the cathode circuit and is shifted by capacitor C5 in order to transmit audio frequencies. The load at the anode end of this triode are telephones which are connected directly to the circuit at the anode. The phones are shifted by the capacitor C7.

How To Make Receiver Parts

Create your own receiver coils. They must be built by dividing them into two pieces to be used on the 40 and 20 meter amateur bands. Each coil should be hung onto carbonite bases made from lamp bulbs that are no longer in use, such as UO-186 or VO-

188 as well as some others. The base's diameter is 38 millimeters. Clear the lamp's bases of any remnants of glass or Mastic, which is the substance with which the lamp is attached on the base. To make Coil # 1 (for 40m band) Take 0.8mm enameled wire, and then wind 19 turns, tapped beginning with the 7th turn. Then, count starting from the grounded end.

Place the ends and beginning of the coil within the base via the holes you drill into the wall inside the base. Then, you can pass both ends of the wire through holes on the lamp's legs. base. Then, remove the insulation and then solder the legs to them. If the wire is unable to pass through the holes of the legs, you can drill holes into the base of the plinth adjacent to the legs.

Solder the end and beginning of the coil on the lamp's legs base, as in the figure. 3. Solder taps connect on the coil for where the wire near the point of soldering should be snipped of any insulation. To precisely determine the location of the tap to the coil while adjusting your receiver's settings, wire must be stripped in 3 or 2 turns. Solder the tap onto the anode that is the lamp's base and then pass the wire via the outside as it's hard to move through the within the lamp base.

Alongside the training for listening it is essential to be able to master the amateur "radio code" the tables that distribute calls of amateur stations based on country, as well as the regulations for conducting radio communication in amateur radio, etc.

Additionally it is important be taking care of the shortwave receiver. It is a fact that traditional broadcast receivers are not suited to receive radiotelegraph stations. If you have a similar receiver (for instance,"Minsk," "Ural," "Record, "Minsk," "Ural," "Rodina," "Record") it is possible to hear a tiny number of amateur shortwave radio stations that operate as a radiotelephone. However, receiving the telegraph signal would require modification in the design of your receiver. It's much simpler to create a specific shortwave receiver. The first step is to select the simplest circuit possible, which means it is easy to build and put up. By using this device, you are able to learn about the shortwaves' operation.

Receiver Circuit

The receiver's circuit is basic and has a twin lamp that is powered by the AC mains and an rectifier. It operates on 20- and 40-meter amateur bands. The transition between bands will be accomplished by switching the coil. The receiver is controlled by an M-type 6N9 lamp. Within the balloon of the lamp, there are two totally identical independent three-electrode lamp and that's why it's referred to as the double triode. The first lamp triode is able to receive signals and detects them. the second is a frequency amplifier.

With only one light source this receiver is similar in size to a receiver with two tubes. The oscillatory circuit in the receiver is comprised of a self-induction loop L as well as a variable capacitor C2. The antenna is linked to the circuit's oscillatory function via the capacitor C1. This capacitor weakens an connection between the antenna and the oscillatory circuit of the receiver since it is connected to a strong connection (for instance, if the antenna is directly connected with the circuit) it is the case that the attenuation of the circuit gets stronger, and the receiver ceases to function.

Additionally the capacitor C1 decreases the influence of the antenna's capacitance the tuning of the receiver. Therefore in the event that the receiver is built precisely as stated, regardless of the antenna used, amateur bands are not going to leave the scale of the receiver. The process of winding the coil is to be performed with great care from coil to coil. As the coil is winding, it is pulled as tightly as is possible in order to stop the coil from sliding.

Coil No. 2. (for the 20-meter distance) is wound using the diameter of 1.0 millimeters, and also with enamel insulation. It is recommended that you wound nine turns using an arc starting from the third turn and counting back at the end that is grounded. Solder the three leads together onto the legs, in the exact order like coil no. 1. Following the the winding coil no. 2, place a long thread between turns leaving a space between turns that is 0.3-0.4 millimeters.

Set up a traditional five-pin lamp socket to the chassis of your receiver, in addition to the sockets that are part of the circuitry for the receiver. In putting the coil in the socket, you are able to include it within the circuit at the three sides. This type of coil that can be interchanged permits you to change from one frequency to the next rapidly. The design can also benefit an amateur radio user in that it allows you to design and test one of the coils first before attempting to make another. Additionally, it's possible to create coils to work with other amateur bands.

A tuning capacitor can be created by any variable capacitor comprising two plates fixed that have a distance of 7 millimeters between them and a movable capacitor must be kept. All the plates that are not removed. The highest capacitance for the capacitor would be 20-25 pF. The minimum value is around ten piF. If you have a capacitor with this capacitance, that capacitor, range of amateur stations can be "stretched" by 15 to 20 degrees Tuning to amateur stations is possible by using an ordinary pen and not using a vernier. If you are overhauling the capacitor, take care to cleanse all the contacts (especially contacts that are rubbing) to remove oxide and dirt and set the rotor's speed to be able to rotate smoothly and easily.

It is also possible to put your own capacitor into the receiver by making it out of two plates, one fixed and one moveable. The capacitor's base is made from natural glass, textiles or ebonite. This plate should be fixed the plate that is fixed to the condenser by using two bolts and a telephone connector, and place it on top of the cutouts in the plate fixed. Install the plate onto one plug. Place the plug in the socket so that the gap between two capacitor plates are approximately 2 millimeters. Then, extend the handle of the plug by putting in an ebonite or wooden stick that is displayed in the front of the receiver's panel. Connect the capacitor using two screws on the vertical panel that is on top of the chassis.

The capacitor plates should be built from aluminum or brass, 0.5 mm thick. As the movable plate is linked to the circuit via an oblique contact between the plug as well as the socket, loud crackling or sound can be heard through phones when the capacitor is rotating. Attach the plate that is movable to the socket by using the flexible copper conductor, or ribbon in order to avoid the negative effects of friction contacts. The capacitor for the antenna is crucial for functioning normally that the receiver. The capacitor's capacitance must be minimal (5-10 pF). It is recommended to utilize semi-variable capacitors for the purpose.

Atop a piece of 1.5 millimeter wire, with enamel insulation and wires connected to the diagram at the top end of the coil. Apply two layers of tissue paper. Then, wind in a second wire having the diameter of 0.3-0.5 millimeters in double paper or silk insulation. Winding length should not exceed 8-10 millimeters.

Connect one part of this cable to an "antenna" terminal. Then, the remaining wire is left unconnected. The two conductors, separated by layers of insulation, make up a capacitor. The capacitance is adjustable by removing or winding turns on an extremely thin wire.

Other receiver parts are factory-made; they need to be ordered in a ready-made. Resistance R2 is variable and Mastic. Its range of value is between 2-3 thousand and 10-15 thousand Ohms. If you purchase this resistance, make sure to check the engine. It needs to provide a smooth, comfortable journey.

For any circuit, which includes our receiver, there are components that must be observed with a particular attention to the electrical properties. On the other hand, some parts are able to be substituted with components with a size that is suitable, but without impacting performance of the receiver.

For instance the receiver we have, these specifics can be different the range of R1 = 1 - 2 mg R3 = 1000-1500 ohm R4 = 0.1-0.5 mg. C3 has 50-100 pF. C5 and are averaging 10 to 100 thousand pF. C4 is 500-1,000 pF. C7 is between 1,000 and 5,000 pF.

It's not recommended to alter the information of coils' turns and wires as well as the way they are wound and frame's size as this can help searching for amateur stations as you set up your receiver. In the event that you do, your receiver will be tuned to signals which are not the same as amateurs. To locate the bands you want then you'll need change the coils' winding to find them, which can take some time.

In the above receiver it is a lamp of 5N9M is utilized. The receiver also was test on the 6N8M (6S N7) kind lamp, which matches the pinout of the lamp. With a lamp of the 6H8M type, the receiver operates without adjustments to the circuit but the outcomes are more shaky. Best results can be achieved using this lamp when the R3 bias resistance is decreased to 500 ohms. If you switch to a 6H8M lamp the ranges of amateurs will be slightly moved toward the right side of the range of the tuning scale.

Receiver Design

For mounting the parts of your receiver it is necessary to build an chassis. To make this happen the best option is to use sheet aluminum that has a 2 to 3 millimeter thickness. Create the vertical panel of aluminum 2mm thick as well as the horizontal one out of aluminum that is thinner (1mm). Attach the horizontal panel to the vertical panel using 3 or 4 bolts as well as nuts.

The horizontal panels on the chassis are on the horizontal panel: a tuning capacitor socket for a lamp with 8 pins, to hold the coil on the vertical side - the antenna clamp with variable resistance as well as telephone Jacks. From the front you can pull out the axis for the tuning capacitor as well as place a knob (limb) that has a 70 to 80 millimeter diameter around the axis. By using such a large knob, even the most weak stations are able to be adjusted (with the right skills) with no particular delay device. Chassis width of 120mm.

The chassis is made out of brass, copper or iron, if aluminum is not available. The horizontal panels can be comprised of plywood or wood or acetate, and then glued with an underlying frame made of a punch-punched capacitor for microfarads. After you have finished the plank made of wood make a bend in the frame, then apply it with a firm pressure to the chassis's vertical panel. The vertical frame should be constructed of metal (aluminum or brass) because a wood panel coated with steel will protect the receiver from the rigors from the hands of the user, and the tuning of the receiver is quite difficult.

It's a good idea to put the receiver on a diameter of 0.8-1.5 millimeters. The best choice is to utilize the wire that has insulation (no no matter the). Connect the conductors using heating the solder with Tin. Make sure you are careful when connecting lamps to the circuit. The socket's petal should match the electrode that you wish to use.

Equipment Setup

It is likely that you have constructed or bought an transceiver or transmitter. The need for tuning it to an appropriate ballast load is apparent and well-known to all. Yet, either way one day, you'll require to pair the transmitter to a particular antenna, and the sending of the tuning signal into the air will be essential. In order to do that you should pick the time to ensure that at present it is clear that the transmission on this band is not too much, i.e., this could be, for instance, time of day for the 80m band and nighttime or close to it in the 10m band, and so on. It is important to note that the two recommendations above are not a law since in the case of, say, just prior to sunset on the 80m band an ultra-long passage may be seen, while during the night, in the 10m band in the times that have the highest solar activity it can be extremely powerful. Most important is to be aware of the frequency you'll be using before turning on your radio. Be sure to remember: "Listen, listen, listen, then continue to listen.

Furthermore, it is your legal obligation to not tune your radio in the DX segment of bands. Also when you begin to tune your transmitter, just according to our agreement, you must listen closely to the frequency. You can check if the frequency appears to be active ("Is this frequency being used?" in English and "QRL?" by telegraph). When tuning the transmitter make sure to check your frequency periodically: the signal might change, and it could be a possibility to disrupt other stations operating on a recently-free frequency. Once you've tuned your receiver to an antenna, you should mark the area adjacent to the knobs to ensure that you don't need to reconfigure your transmitter with every change between bands.

To become a radio amateur, you must get an amateur radio license. It usually goes like that you submit your application to the nation's amateur radio association as well as directly with the administration of communications in your nation, and complete an exam to qualify. exam and yes it is necessary to show that you're prepared to receive the privilege of using radio for a leisure activity (and not in commercial use in exchange for the sake of money) Based on your success in the exam you receive a certificate as well as a call sign. Call sign: the only and sole identification of the radio enthusiast that identifies him with other amateur radio users. the same callsign does have no place; they are all distinct.

Ham Radio Call signs

The design of amateur callsigns is simple and straightforward, however it is not without exceptions that we won't dwell on. The typical look of a callsign is similar to AB1CDE.

AB can be a suffix with that a particular territory can be identifiable (a country, or a specific part of a country when a specific division is defined). The prefix typically consists of two letters. Sometimes, it's a word, but sometimes it is a numeral, or a letter. It's easy for people who are not fascinated by aviation to discover the prefixes for amateur radio. They correspond to ICAO prefixes.

1- one digit. This is an integral part of the amateur callsign and is also used for distinguishing it from an aircraft's call sign. The callsign's number differentiates the prefix and the suffix. In the event that the prefix is one letter and number. In this scenario, the next digit following the prefix may be omitted but only in extreme circumstances (memorial the callsign RAEM) is that the digit included. The majority of time it is more than one digit most of the time a temporary call sign to commemorate the occasion, or a specific time or date. Some countries have numbers are not meaningful and it is released in sequence of numbers increasing or random. Certain countries' number is a reference to a particular geographical region. In certain countries, it refers to the type of license, or additional specific information.

CDEA suffix composed of one-to four letters. Usually, the letters are dominant in the order of increasing numbers. The first letter or the second letter may carry a significance for instance, a geographic area of a nation or the name of a license class. It is not necessary to learn all that information; it is important to know the fundamentals; commonly occurring calls will be remembered for themselves. However, unique ones, which are obviously unfamiliar to you, may trigger an insatiable urge to "work" at the station by introducing an entirely new sign quickly as is possible. For instance, consider the W5UN call number - an actual operator who is well-known within his circle. W is the word for continent of the United States; there could be a single letter or two however this particular operator uses one. 5 is a conditional area (AR, LA, MS, NM, OK, TX) as well as the letters unreceived are can be used in succession or chosen among the ones that are not occupied. The conditional area isn't assured since within the United States, it is permissible to use your phone number when moving. You can also be given a phone number that you prefer (from no cost ones) and include the digit of a different region. Also, we can assume from only four digits of the call number, however this is only a case that is only applicable to those in the United States, that the provider has the top license class i.e. an absolute tolerance. It is the case that in many nations (but in some, like in Japan) whenever you upgrade the class of your license it is possible to exercise the choice of an easier call sign. But, not every person has this privilege, therefore an extended call sign will not always mean you have a lower level of license, however the short call sign almost always indicates a top-quality license.

Let's look at a different example: P3X is also a live operator. Through the prefix "P3X" we identify Cyprus and the country's name; however, the prefix is usually omitted and the suffix has only one letter, X. It is a unique callsign to be used during competitions. It was given to an radio amateur using the long normal callsign of 5B4AMM, only for the radio sport competitions (we'll return to events later) in which 5B is the prefix, while 4 is the number that separates from the prefix, and AMM represents the suffix to the name received.

What Exactly Do Radio Amateurs Do?

The principal directions include:

1. The design and manufacturing of equipment and antennas. Radio stations of today are an extremely complex piece of equipment both in its design and manufacturing which is why the industry provides various kits with different levels of sophistication which allows people with different degrees of experience to build their radios and not leave it to engineers who are independent. The majority of people who own industrial radios make simpler devices just because they enjoy it, and not simply due to the fact that they cannot purchase factory equipment. This is also true for ancillary equipment and antennas that are made to many isn't an opportunity to cut costs however a means to express their passion for this craft. However, it is nice to create an association with your own device, even though your product isn't as great as manufactured equipment by an order of the magnitude!

2. Sport. There are a variety of radiosport events that have a common theme that is to create as many QSOs as you can in a set time as well as conditional points are awarded to for each QSO. The more points you earn, the more difficult and technical it becomes. There's not much time to talk about this issue and only solve the fundamental issue of connecting - that is, the exchange of signals and reports about the strength of the signal and, in general serial numbers, according to which refereeing takes place. There are a variety of award schemes where operators are not restricted to time and the time frame can be very broad It is required to create a set quantity of QSOs in accordance with the regulations of each award to be awarded an official certificate of recognition or a plate that confirms the accomplishment. In order to fulfill the requirements for the award you must join more than one hundred "territories" in the DXCC list. Currently there are the number of 340. The majority of the

time, each administrative nation has the equivalent of one DXCC territory.

3. Talk about radios, antennas, as well as other related topics. As the radio topic is a common topic for everyone who listens to radio, the conversations are centered around radio. Some countries have negotiations that are strictly restricted, including antennas and radio equipment radio-related sports, radio wave propagation, as well as weather. Certain countries don't have any specific restrictions, however the rule of law remains the identical. In some countries, there's a ban immediately on the transmission of information to third parties even if they're radio amateurs. If you're uninterested in the subject then you don't have anything to have to do with radio amateurs since they do not have any other subjects to discuss. Religion, politics, commerce as well as offensive behavior are all strictly forbidden. If you are not part of

the subjects that radio amateurs are allowed to discuss, there's a CB band,

How Are Connections Made?

There are a few kinds of modulation. Most of the time, they are in sync with each other, however it is also possible to connect them using different ways. Communications are usually conducted at the same time, although a tiny interval of time and also various ranges of frequencies are commonly utilized. There is no one with their own frequency. This frequency, which is not currently occupied, is temporarily used by an operator who is making a general or directed phone call. If anyone hears his voice and is willing to respond - does. When the call operator has finished his work or switches the frequency, the active frequency will be let go, and any other person will be able to immediately use the frequency. First, the operator calls the number of whom he's calling or makes a general phone call and then calls his personal call sign. The person answering his call starts by dialing a phone number, the one that he calls after which he calls his own.

Options For Signal Modulation

When it comes to signal modulation choices are available. Telegraph and Morse code. Radio amateurs typically call it CW but it may be better to refer to it as A1A. The reason why radio amateurs who, at their existence were at the forefront of signalers, with numerous achievements when it comes to mastering the art of radio communications but then hit the reverse track and employ an old-fashioned method of modulation that signalists from professional organizations have long since left, and in a lot of nations, the capability to transmit via hand and also receive is voice telegraph still an element of the required exam? From the perspective of the administrations for communications of various countries might sound like this: as the amateur radio channel in times of emergency are a viable alternative to commercial or state channels which were shut down due to some reason and it's essential to have the ability communicate with each other, even on homemade equipment made from just one transistor, as the lives of others could be at risk.

However, among radio amateurs themselves, the opinions are divided. Some claim that it is the most "long-range" kind of radio, but this isn't the case ("digit" does not mean "long-range"). A person is not technologically advanced and works with notions such as "we learned - and let them (beginners) teach." If you're an amateur radio is an activity because every game has a set of conditional rules. The radio players also prefer to adhere to these rules. The main issue with the telegraph lies in the necessity to understand transmission (not extremely difficult) as well as reception (more challenging, yet difficult, but not impossible). Additionally, it has a an extremely low rate of data and is not an excellent view for sports reasons, even with its retrograde.

Types of Digital Telephony

The vast majority of it is majority of it owned by SSB (J3E). Note that equipment designed to work with SSB is quite complex to construct and install thus the move to SSB was delayed for quite a lengthy time while remaining loyal to the AM (A3E). With the development of equipment from commercial manufacturers and the release of reasonably inexpensive amateur innovations and the introduction of SSB succeeded in achieving the undisputed victory of SSB that provides roughly 50% of the bandwidth occupied that transmits the same amount of information which means that it saves radio frequency resource and it has the potential to increase the ratio of signal to noise. Nowadays, AM is almost never used. Second most used with a huge margin can be found in FM (F3E) that is more popular for amateur VHF bands as well as non-amateur CB band. There are a variety of digital telephony where operators' voice is digitalized and sent as a digital stream.

At present, these choices are considered to be somewhat in the beginning, however their importance will likely increase significantly in the near future. While it appears to be simple connecting in telephony, it's even more simple than using a microphone and shouting - this is actually one of the more difficult methods because it has a higher ratio of signal-to-noise than the telegraph and higher than other varieties of "numbers."

Chapter 3: Running Your Ham Radio Station

We will briefly discuss the operation of a radio station (synonym"transceiver"). Transceiver, Translated from English refers to an transceiver. (Transmit Transmitter - transmit, Receive - get Transmitter - transceiver). According to the description that the transceiver is composed of two components: the transmitter and receiver. The transmitter and the receiver are both complex and hidden within a housing that is populated with nodes. The job for the receiver radio is converting the radio frequency signal that it receives from radio frequency ether to the sound that is perceived by the ear of the person. The modern radio signal receivers utilize superheterodyne schemes (with dual the frequency of conversion).

A radio station begins with an antenna connector. The signal that enters the connector goes through a bandpass filtration system and amplified with a high-frequency UHF amplifier. After that, the signal gets sent to the detector in the radio station. The detector extracts low-frequency information part (for instance the voice) in the sound. After the output from this detector, the sound goes to into the amplifier for low frequencies (ULF) followed by the speaker. The ULF converts radio signals transmitted through the antenna connector to an audio sound signal that can be heard.

The quality of radio reception is contingent on the performance of each of the mentioned nodes. It is for this reason that some radio stations are heard clear and loudly, while others receive with interference and unclear, mumbling. The reason for low reception for the radio station could have to do with the incorrect installation of the antenna of the radio station operating at 27 MHz. Radio stations' quality offered to consumers is always shifting, so in selecting a station, it is recommended take the advice of experts.

The Transmitter In A Radio Station

The function of the transmitter on the walkie-talkie performs the same function of that of the receiver, which is to relay information from one subscriber to the next. The transmitter in the radio station, there is a procedure that almost exactly the same as what happens in the receiver. The information (for radio stations it typically consists of voice however, it could be data as well) is layered on top of the frequency of the carrier set through the generator of frequency, and transferred via cable to an antenna in the air. If we send information to an radio station via the voice. The audio signal received from the radio mic is sent to ULF (low-frequency amplifier) that is located following the receiver. There, the signal is amplified, then transferred to the modulator.

Modulator is a component of an radio station that alters the frequency of high-frequency (HF) radio signals that is broadcast through the air in accordance with the law that alters an information message (voice). For radio stations operating at 27 MHz amplified modulation (AM) and frequency modulation (FM or FM) are the most frequently used. Amplitude modulation (AM) in a walkie talkie involves a change in the intensity of a signal with a high frequency according to a law that governs voice changes. When the final signal is transmitted through the air, frequency of the signal remains constant however the amplitude fluctuates. On the other end the RF signal's envelope will be assigned. This envelope is known as the signal that contains information.

The frequency modulation of the radio station refers to a variation in the frequency of the radio signal that is broadcast over the air based on the laws of voice change. In this instance the RF signal is the same amplitude and the frequency of transmission varies. On the receiving side of the FM detector, it picks the signal with low frequency and transmits it through the ULF before feeding it into the speaker. It also has one-sideband SSB modulation (USB as well as LSB) with more complicated algorithms and is integrated into expensive equipment. This is why it's not commonly used by the 27 MHz band of radio stations.

Step By Step Setting Of Ham Radio Transverter

It is suggested to begin the process of setting up a design with a thorough inspection of the proper installation. Beginning with tuning should be done using the use of a quartz oscillator. It is important to connect to the base of the 1T5 transistor to the case by using an RC capacitor that has a range of between 1000 and 5000 per centimeters. This way the quartz oscillator will transform into a standard LC oscillator.

The production frequency is controlled by the circuit 1L91C19 1C20.

In order to rotate the core of 1L9's coil it is important to establish the frequency that is near to that of the three-frequency quartz Resonator. Following that, the blocking capacitor will be removed from the base part of the 1T5 transistor and then fine-tuned until it is at an appropriate position where the rotating of the core of the IL9 coil will have the lowest impact on the frequency of generation.

If you have an electronic receiver that has an amplitude scale, or more importantly and more reliable, an electronic count frequency meter, the production frequency must be assessed and, if required and adjusted. The fact is that the methods of correction using electricity are not effective for circuits that operate with mechanical harmonics in an resonator made of quartz. Thus, the only option is to modify what parameters the resonator has. This is the case for a resonator made of quartz that is equipped with the use of external plates made of metal, i.e., without metallization of the quartz plates. The frequency of this resonance can easily be enhanced by 35% through grinding the quartz plate with smooth Sandpaper.

The frequency of this Resonator may be reduced to 0.5 percent of its nominal value by simply rubbing the center of the plate using an ounce of lead or solder. When doing this you should take into consideration that the plate treated this manner is susceptible to becoming aged within the next 2-3 days. Following that time, the frequency changes stop while the quartz resonance operates smoothly. It's much harder to alter the frequency of quartz that has been plated with metal plates. If plating is made by using silver, the resonator's amplitude could be enhanced by decreasing the coating's thickness with an eraser made of ink. If you want a stronger coating, make use of a fine-grained abrasive.

In order to turn to the quartz resonator inside the circuit, it's important to wash the plate using a damp cloth that has been soaked in alcohol. After that, install the multipliers' chain that is part of the heterodyne pathway. In creating the multipliers as well as any other steps of the transverter it's essential to regulate the operation modes of the transistors that operate in direct current. It is more convenient to determine the voltage on the collector as, when there is the known resistance to the resistor inside the circuit for collecting, it's easy to calculate the amount of flow of current through the transistor. = (Ep Ek) (Rk) where I represents the flow of current across the transistor. MA, Ep, voltage of the power supply V; Ek is the voltage at the circuit's collector, V. Rk is the collector's resistance resistor. It is in kOhm.

One of the features of the measurement mode is that it must be conducted in an operational state, that is, when there is signals. In reality, the majority of the transistors utilized within the radio station work in a large signal mode meaning that the operational modes for direct current as well as high frequency are linked. When this happens, the connection of the probe with the measurement device could interfere with the functioning of the cascade the high frequency, and consequently cause an error to the results. The other risk is that in the case of measuring the signal mode of a transistor that operates in the low signals mode, the cacade could be self-excited if the probe is attached.

This self-excitation could significantly alter the performance of the transistor, and consequently affect the measurements. To minimize the effects of self-excitation the user must conduct measurements using an resistor that has an resistance of 10 kOhm and above. The resistor should be placed near the tip of the probe in a way that the conductor attached to the circuit is of an appropriate length. A second resistor can cause a reading to be understated by the voltmeter however the error is simple to determine. In order to make your measurements easier it is possible to choose, for example, to switch to a lower level of the Voltmeter. After you have selected the resistivity that an outside resistor provides switch back to the earlier scale.

The creation of the initial triple, which is made using the 1T6 transistor, starts by adjusting the excitation mode. When selecting the capacitance of 1C22 capacitor 1C22 capacitor, it's important to make sure that the constant voltage at the circuit's collector is between 5 and 6 V. This equates to the current at the collector for the 1T6 transistor, which is around six milliamps. Then, they begin making the dual-circuit filter 1L10 1C25-1C26. The settings are made by adjusting the highest collector current of the transistor/G7. It will be the second stage that the multiplier. The amount of excitation required that is required by the 1T7 transistor can be altered by adjusting the connecting point of the filter circuits between both the collector part of the IT6 transistor as well as at the bottom of the 1T7 transistor.

Be cautious when choosing the taps for coils to make sure that the circuits are able to be loaded equal. The load quality of the circuit may be determined by the clarity of the settings using an adjustable capacitor called a trimmer. one circuit is having an oblique or "blunt" settings, the coil's tap must be connected closer to the terminal that is grounded. When the proper setting is used that the constant voltage inside the collector of the transistor 1T7 is 5-6 V.

If the sizes of the coils 1L10 and 1L11 are precise as well as the capacitors for trimming are mid-way between them it is not a risk of tuning the filter into the incorrect harmonic. But, in the event that the sizes of the coils or speed of crystal oscillators have been modified, it's beneficial in some way or another, to determine whether the settings are correct. It is possible to, for instance make use of a receiver within the frequency range you want to operate. A wire piece must attach to the receiver's input and the other end is connected into the circuit 1L10 1C25. While moving the tuning capacitor 1C25 the highest volume of signal has to be equal to the current at which it can collect. the transistor 1T7.

Testing Your Setup

The options for this testing procedure are constrained due to the fact that many communications receivers operate with a frequency of less than 25 MHz. It is possible to increase the spectrum of frequencies received by the simplest set-top boxes. It is a quartz self-oscillator based by GT311 transistor. Additionally, time the transistor is able to perform the function as a mixer that operates using the frequency harmonics of the quartz self-oscillator. In order to do this it connects the oscillator to the shortwave input on a receiver by a cable.

In order to establish a heterodyne pathway the prefix should be linked to the circuit of the adjustable multiplier by using a small piece of wiring. In order to do that, connect the end that is insulated of the wire into the "hot" circuit's output. There aren't any selective circuits within the prefix. It is because reception happens in a continuous manner on the various harmonics of the oscillator. This aids in understanding the range of signals are generated by the crystal oscillator of the local oscillator as well as the crystal oscillator in the set-top box are already known. The attachment could make use of any quartz resonator that has the natural frequency of 8-15 MHz.

Consider, for instance, how to tune to tune the 1L10 1C25 circuit up to 61.5 MHz. The set-top box will employ a quartz-based resonator to 9620 kHz. After that, a check of that the oscillator in the transverter's crystal shows it has a frequency of 20504 KHz. The signal is heard through the fourth and fifth harmonics of the oscillator local to the box. In the first instance you should looked for using 61 512-9620-4=23 032 kHz. If you are in the second situation one, that is suitable for receivers having a narrower operating area The signal must be sought with the frequency of 61512 + 9620*5 = 1312 KHz.

Expanding The Frequencies

You can control the proper tuning of multipliers to up to 400 and 500 MHz frequency. In general, the frequency range is able to be increased by using a higher frequency transistor employed as well as the capacitance capacitors C2 as well as C4 are reduced. Correct setting of multipliers may also be verified using a resonant wavemeter, and, more ideally, the spectrum analyzer.

We will continue to look at an approach to adjust an oscillator local to an oscillator 144/21 MHz. When the excitation required is placed on at the bottom of the IT7 transistor, the circuit begins to tune the 1L12/1C30 circuit at an amplitude of the 123 milliseconds. The second cascade that follows the doubler is an amplifier that uses the 1T8 transistor that operates in the class A mode. Because of this, the current at the collector from the IT8 transistor is not dependent upon the voltage of excitation and does not serve as an indicator for the settings of 1L12's circuit for the 1C30 doubler.

So, it is essential to tune performed using a receiver or, in a simpler instance, a high-frequency device attached with the test.

The tester needs to be turned towards the highest sensible DC measurement range. The level of connectivity to the circuit may be altered by shifting the location of connection to the line or the coil.

When that the 1L12 1C30 circuit has been adjusted to the frequency desired it is then tuned to the terminal amplifier on the heterodyne circuit on the 1T8 transistor. The first step is choosing that resistance on 1R20, the 1R20 resistor, it's important to establish the current collector for the 1T8 transistor in the range of 7-8mA. This selection should be done without an excitation signal. Following that, the 1T8 transistor needs to be stimulated with an ultra-high frequency probe, you can set up 1L13's 1L13 1C34 circuit. The local oscillator is now complete. set-up.

Setting Up The Receiving Path

The process of setting up the receiver route must start by selecting the transistors' modes 1T9 and1T10 for direct current. When selecting the resistors 1R22 and 1R26 The collector currents of transistors must be kept to be within 2-2.5 milliamps. The mixer should be connected to the input of a shortwave transmitter that is tuned to the frequency of 21.2 milliseconds, and the circuit 1L18 1C50 1C51 is tuned for the highest frequency. Following that, the high-frequency probe should be connected to circuits 1L17 1C45 the next step is to connect 1L16 1C43 and the bandpass filter is adjusted to maximize the oscillator's local signal. After that, by gradually decreasing the power that the capacitors for trimming alter the bandpass filter until it is at the highest signal. The adjustment process ensures that UHF does not tune in the direction of mirror channels.

The circuit for input IL15 1C39 must be equipped with an input signal. The input signal may be, for example, the fifth harmonic from an RF transmitter within the region between 28-29.7 MHz. In order to do that, switch the converter's input to a shunt using the 75-ohm resistance and attach the wire to a length of 15 to 20 centimeters in length as an antenna. Also, you can try receiving signals from radio stations that are two meters. It is however more practical to make use of the noise signal source as it is, in this instance, the process of tuning isn't dependent on the fluctuation of the level and frequency of the signal. The 2D2S tube noise diode could be utilized as source. The main benefit of this particular source is the fact that it produces an amount of noise that is known which can be utilized to determine the noise level of the receiver. However, the drawbacks are the fact that the highest noise volume from such sources is very low (20-50 kT0) and, in addition that the higher the intensity of the noise it is, the more cathode's temperature, and consequently lower the diode's time of service.

In this regard, it's best to use the noise diode for ultimate tuning process of the receiver. You can utilize, for instance the use of a diode semiconductor noise generator for the initial tuning. This kind of probe can be seen in Fig. 27. The source of the noise can be traced to the emission junction of transistor KT306 operating in reverse break-down mode. In this instance, the magnitude of the noise can be as high as several hundred KT0. This allows you to boost this SWR for the instrument through adding an attenuator to resistors R2 & R3 that have an attenuation factor of 13 dB for the output. The probe is placed in a tiny box that has an extension cable that connects it to the receiver's input.

During the installation process, focus should be given to the length that is minimal for the wires connecting the transistor 77, as well as resistors R2, R3 and the capacitor C2. It is particularly important when the probe will be used for tuning 432/21 and 1296/144 transverters at MHz. This probe produced a great performance with the GA402 germanium microwave diode. It has lower capacitance and inductance of the lead, that is particularly important in higher frequencies. It is easier to adjust the probe to set the maximum current that flows that flows through the diode to 1-3mA. In order to ensure stability the voltage from the power source is required to be at least two times more than the voltage that the disintegration of the diode occurs. The voltage is controlled through the selection of the resistance of the resistor R1.

By using this device, you are able to easily adjust the receiver pathway to achieve maximum gains. For this it's necessary to connect this tester with the input of your main receiver and switch it to the AC voltage measurement mode. Adjusting the circuits, and choosing interstage connections will give you the best readings. Final adjustments are made with a measurement noise generator. This method of adjustment will be discussed in the following.

Setting Up The Transmitting Path

You can then proceed to create the transmission pathway. In the beginning, you must select the transistor mode for direct current. When you select the 1R10 resistor The voltage at one of the 2T4 transistors' collectors can be set to +7 V which is equivalent to an i.e. 10mA of current. By using the resistor 1R8 the transistor 1TZ mode can be set. The collector's voltage 1TZ must be equal to the value -I) (collector current 20 mA). (collector current 20mA). In adjusting the initial current for the terminal as well as terminal transistors it is more accurate to determine the DC voltage that is on the collector, not in relation to ground, but rather in relation to positively wire.

A voltage drop on the resistor 1R4 must be at least 4 V (100 milliamps) The current drop in the voltage across the IRt resistor must not exceed 0.2 V (40 million mA). Then, the supply voltage of transistors 1T1 as well as 1T2 has to be switched off. You can now begin to tune the resonance circuits. Initial tuning takes place using a signal that has the frequency 21 MHz. In this instance, the resonant circuits 1L81C15, 1L7, 1C14 and 1L6 1C10 have been tuned to the frequency of the local oscillator, i.e., to the frequency of 123 millimeters. Tuning can be done by using a high-frequency probe that is connected to the circuits. A signal that has an amplitude of 21.2 MHz should be fed to the mixer's input.

The signal voltage has to be raised until a significant drop in the collector voltage in the 1T4 transistor is observed. In the meantime, circuit 1C35, 1L14, and 1C37 are adjusted. time circuit 1L14 and 1C35 and 1C37 can be altered. The signal from the local oscillator at the mixer's output will be reduced. The high-frequency probe needs to be connected to the 1L8 resonance and then, through the rotation of the 1C15 tuning capacitor with decreasing capacitance, locate the closest voltage peak that corresponds to the frequency that is 144.2 MHz.

You can now configure the two final steps of the transmission route. In order to prevent malfunction of the transistor/ and the output of the transmission path has to be connected to a source that is proportional to the impedance in the feeder. The load is created independently by connecting a number of in parallel two-watt resistors that are of the MLT variety. This could be as simple as 4300 ohm resistors when the feeder has a typical impedance of 75 ohms has to be utilized as well as six 300-ohm resistors in the event that the resistance of the feeder of 50 Ohms. The load resistors along with the detector are put inside a tiny metal box that has a high frequency connector. Resistors R1-R4 should be arranged in a Star pattern with respect to the connector, and must be of a minimal length. If the detector comes with a dial indicator it will be a standalone device that is the most simple power measurement device. It is recommended to add an electronic switch to alter its resistance to the resistor which will affect the measurement limits of power.

Once the load has been hooked up to the output on the transmitter route and the voltage of supply applies to the remaining two stages, the circuit is set to begin making an 1LA 1C6 circuit. It is set in accordance with the current at which collectors can be set for the transistor 1T1. Before that, the transistor 171 has to be hooked up to a load as far as it is feasible, i.e. you must set the capacitor 1C1 up to maximum, and the capacitor 1C2 to lowest.

The collector current of a 1T1 transistor may exceed 500 mA and even more. If the voltage of excitation is not sufficient, it's important to set all the stages that are in the beginning and reduce slightly the capacitances of capacitors 1C5 and1C7. The output circuit can be adjusted in accordance with the maximum reading displayed by the indicator of power. The higher the capacitance one of the capacitors 1C2 weaker connections to the load. When the connection is weak, and an excitation limit that is too high, the transistor may enter an overvoltage high mode that is, there's an increased risk that the transistor could be out of alignment. This is why such situations should be kept out of.

Selection Of Best Equipment For Your Ham Radio Station

It's also crucial to consider which location we intend to place the station. The station will not be as effective on the field, which is normally without interference from neighbors, like in cities, which has a greater radioelectric hazard. As we look at the latest developments of radio enthusiasts it is possible to say that all of us want an affordable device that can cover the majority of possibilities we have for conducting radio. This is why the popularity of multimode (FM, CCW, SSB) and multiband (HF VHF, UHF) equipment has. If we are looking to establish the station which allows us to " end the itch " (do a little DX with HF, make use of digital modes, maintain QSOs with colleagues who are usually in the 80- and 40-meter range as well as use UHF and VHF repeaters, test satellite communications, use radio packets, APRS, etc.) If you're looking to establish a station that is without doubt, a great choice could be to join one of the teams listed above.

Comparison Of Equipment From Different Brands

Five teams have been selected in response to what I have previously mentioned that they belong to three major brands of equipment producers that cater to radio amateurs. Kenwood, Yaesu, and ICOM. It is important to note that the comparison is solely didactic and is not based on any commercial interest towards each brand.

There are two of these, FT-847 as well as the TS-2000 could be considered base equipment. And the others, which are mobile or base-mobile.

Then, we will examine them in relation to test results conducted through the ARRL. The ARRL tests must be noted that these tests may be subject to some degree of error because of the calibration of the measurement equipment used and also the calibration of the transceivers as it's almost impossible for them all to be set the exact same way from the factory.

EQUIPMENT TX RX

Sp IMD MDS BDR

3rd 5th 3.5 14 50 144 432 3.5 14

Twenty 5 Diff Twenty 5 Diff

TS-2000 55 27 42 138 137 142 140 143 127 99 28 126 99 27

FT-847 50 28 51 137 136 140 142 141 109 82 27 109 82 27

IC-706 53 30 33 142 142 142 142 143 118 86 32 120 86 3. 4

FT-857 53 25 40 136 137 136 140 140 109 90 19 109 90 19

FT-897 53 23 37 137 137 142 140 139 111 85 26 109 85 26

In the first column, on left are five of the teams we have selected: Three teams from Yaesu and two are from Kenwood and Icom and Icom, respectively. It is no doubt that Yaesu is the longest-running company manufacturer of this kind of equipment and also has the most models on its list. The transmission measurements are different from reception measures within the second division.

When it comes to transmission, we can say:

1. Sp.This is the degree of spuriousness that is present in the transmission signal. It is a measure of the purity and pureness of the signal. It's measured in dB The transmitter is more effective the greater it is in the table.

2. IM D . This is an intermodulation that occurs in the transmission of 5th and 3rd order caused by two similar tones like 2KHz and 1KHz. These are put into the modulator. It's expressed in decibels and the modulator is better the higher it is listed in the table.

We will be having reception at the reception. will have:

1. MDS. The minimum detectable signal to the receiver. It is also known as sensitive. It's measured in dBm which means that the receiver will be more sensitive the higher number on the table. This is an important bit of data if you want to detect weak signals like satellite reception.

2. BDR. This is the dynamic blocking range which is the ability to remove signals that

are adjacent to those we want to receive. It is that is measured at five and twenty KHz. This is directly connected to selectivity of a receptor. This is determined by putting the receiver in a specific frequency, and then placing it next to an additional piece of equipment that can transmit at five or 20KHz greater and lower. The process begins by slowly increasing the output of the equipment until it blocks the reception of the initial signal.

3. IMDDR. This is the dynamic range of distortion due to intermodulation or the capacity of the receiver circuits to not generate false signals even in the presence of two powerful signals coming from the input to the receiver. The measurement is in decibels as well as the receiver's performance is more efficient the higher the number listed in the table for 20 and 5KHz, as well as smaller the differences between the two.

4. IP3 is a third-order intercept feature that many manufactures use to emphasize the receiver's performance. The IP3 is derived from the mathematical formula of the parameters previously mentioned. It's measured in dBm The receiver will be more effective the greater its value.

From the first glimpse it is clear that the TS-2000 is, not surprisingly, one of the highest values across the majority of parameters. It's second to one or two in sensitivity to the IC-706MKIIG. The IC-706MKIIG, however is superior to all Yaesu with respect to most parameters, not just the FT-847. The main differences are between the two teams are that each team has the same internal electronics. The only difference between them is the exterior casing, which is why we discussed at the start concerning the adjustment between both teams.

Then we will do the second analysis of the two teams. This time, we will focus on the electronic components that make up their circuits. The table below gives an understanding of the transistors that are used to transmit data from the device.

Equipment Power Band Final Transistor Drivers Driver Amplifier Preamplifier Start

TS-2000 100 HF 2SC5125x2 2SC1972 x 2 2SC1971 2SK2596 2SK2596

100 VHF 2SC2694 x 2 uPC1678G

70 uhf 2SC3102 2SC3022 2SK2595 2SK2596

FT-847 100 HF 2SC5125x2 MRF5015 MRF9745 uPC1677

50 VHF 2SC5125 PF0310 UPC1677

fifty UHF 2SC3102 PFC0340 UPC1677

FT-897

FT-857 100 HF 2SC5125x2 2SK2975 x 2 2SK2973 x 2 2SK2596 2SC3357 uPC2710

50

20 VHF

UHF 2SC3102

IC-706 100 HF SRFJ7044 x 2 MRF1508 MXR9745RT1 2SK2854 uPC2709

Based on your radio amateur knowledge, you could invest in more expensive equipment. If you're new to the hobby, we suggest that you begin with a simpler equipment for ham radio. Place it at the end of your table, and then have fun with. Then, you can set the entire space aside for the Ham radio. To get started, purchase an amateur radio license. If you want to join the ranks of radio amateurs, it is necessary to be familiar with these aspects:

1. Security. Be sure that your fire alarm system inside your home is correctly wired for the ham radio

2. Fixed antenna. They should be correctly positioned and secured;

3. Radios. It should work properly and connect correctly.

4. Be aware of the way you will set up and set up your radios. The enjoyment you get from your hobbies is dependent on this.

Eliminate Radio Interference With Simple Means

At home Radio interference can be an annoyance for many when the stripes or blocks on the TV image interfere with watching enjoyment, or if strange squabbles or whistling sounds can be heard over the radio or the reception can be weak. Radio interference can be the cause of conflicts between radio amateurs as well as his neighbors. Each party can be in the process of being affected as well as the one who is at fault. It is essential to be educated with a language even people who aren't familiar in radio technologies can comprehend. This article is designed to assist to achieve the need for.

It is possible to help in a majority of instances without having to directly interfere with any malfunctioning or unreliable equipment. In fact, it's not just radios that create interfering. It's usually devices like plug-in power supply, televisions or even computers that inexplicably disrupt the radio reception of shortwave, amateur listeners, and FM radio, without polluters not even knowing about the issue. Certain gadgets produce a continuous the sound of a hissing that makes the radio stations seem ineffective.

We search for a location without reception AM radio, and then make the radio louder. After that, we switch on the lightbulb. The radio squeaks. It's normal for physical physics. The current pulse generates a number of radio waves at diverse frequency. It's what happens in a storm. There is already radio interference and it is that small with a light bulb that it is possible to endure it. Keep in mind that turning on generates a current wave of radio frequencies that are that are radiated by an antenna. When it comes to the bulb that lights up the light's network acts as an antenna.

Radio transmitters first appeared as pop sparks using the same principles. It was possible to use them to connect the Atlantic. However, they were quickly banned due to the fact that they interferred with other radio networks on a high-speed basis. Nowadays, it's predominantly switching power sources that produce the required voltages. are produced by over 30000 switching operations in a second. Thanks to the switch-on switching-off time ratio that the voltage is maintained very efficiently as heavy and costly transformers are not needed for 50 Hz AC voltage. Furthermore, the power loss is comparatively low. Unfortunately, these processes could emit a significant amount of high frequency energy into power lines as well as other lines of supply if appropriate countermeasures aren't taken in relation to circuits and some may are expensive. There are great-designed products that cause minimal or no disturbance, but some products disrupt the system. An insignificant power supply plug-in is one hundred times disruptive that a huge TV you have in your living room.

Type Of Interference

The following could be difficulties between a radio amateur and the radio amateur's neighbors. These can be resolved in most situations with a simple solution:

First case: Radio amateurs feel a lot affected by the interference of electronic devices within the area. They are unable to hear weak radio stations since they are obscured by the sound and the crackle of radiations of the interference. Most of the reasons are inadequately designed radios, televisions as well as power packs, computers as well as other electronic gadgets, that emit high-frequency signals into the the light network, and hinder reception for a distance of as much as hundred meters. In recent times, there are increasing incidents of inverters that cause disruptions within solar panels. Inverters convert DC voltage of the solar cells to the AC voltage that is generated by the grid.

The radio amateur is required to pay a fee annually to access his frequency and is able to contact to the Federal Network Agency for help. If this is the case it is they will be able to help. Federal Network Agency is entitled to operate in private residences in the event of an appointment being set. If someone refuses to work the appointment, they can enforce it using violence in certain cases. If you hear the Federal Network Agency rings at the front door to your home, it's impossible to ignore the door. The agency will return, if required, with the police. It gets costly.

If a radio enthusiast observes that the neighbors do not want to collaborate in removing the interference they will contact to the Federal Network Agency with his skills at some point or other. In general the agency will employ the aid of a measurement vehicle to get at the root of the issue.

2. The radio amateur appears to be interfering. The most common disturbances are those that affect pictures which occur in intervals of a few seconds to minutes. The sound is emitted out of the loudspeaker or telephone. Make contact with the radio amateur, and explain the cause of the interference, and how it happens. This is part of his education to identify and correct these errors. He'll be keen to get to the root of the issue and removing the root cause. In this regard He needs your assistance. Can you insist on him stopping the hobby? Radio amateur is not an activity for a lot of radio enthusiasts. To them, being part of a vast international community represents an exclusive way of living.

Solution: It is interesting to note that the technology solution for both situations is similar for the majority of cases. One of the reasons is that the immunity to interference of disrupted devices needs to be increased. However it is essential that the unwanted high-frequency radiation of the devices that interfere with them must be eliminated. In both situations the frequency is pumped through the devices that are disturbed via the supply lines. Or as in the case of a reverse the interference high-frequency gets released through supply lines. The mains cables and audio cables, speaker cable, antenna cables telephone cables and others are all questioned. They function as an antenna for receiving or transmitting. Filters are placed in these cables to block interference and make sure that there is no radiation in the lines. Soundproofing that is effective in the residential building is a good frequency electronic decoupling. Direct radiation-related disturbances direct radiation - i.e. the absence of cables - are not common.

Finding a Ham Radio Club

Clubs for ham radio have been around since the beginning of radio hams, providing an easy way to communicate with fellow enthusiasts. In the beginning of the ham radio community, they were made up of people who had similar desires and worked to create radios, even though the technology was at its beginnings, and the possibility of it was difficult to predict success.

Ham radio clubs are great resources of assistance and advice to those who are just beginning. When you're new to the ham radio hobby there are bound to be many basic questions that need an answer. It's best to sign up to a group that caters to a broad range of preferences and, in the event that it is it is possible, focuses on providing assistance for new ham radio enthusiasts. A group of people who share the same interests will make your journey to having fun with ham radio a lot easier.

A lot of ham radio users belong to several clubs. This includes general interest clubs and a few specific interest clubs. Local and regional clubs generally organize physical gatherings and attract members who reside in a particular region. The specialized clubs concentrate on specific things, including contesting the low-power competition, operating with less power, or amateur television. They may include members from different areas. Certain clubs conduct their meetings solely through radio broadcasts, instead of in-person gatherings.

Simple online searches can be used to find nearby ham radio clubs as well as you can go to the American Radio Relay League (ARRL) site can be used to find a list of affiliated clubs, based on their the location. Clubs of general interest that provide support to novice hams must be considered first.

If several radio clubs operate in your area, the meetings times must be evaluated with the best choice should be chosen. If you have a particular interest It is recommended to look into whether local clubs give time for them. Going to multiple meetings in different clubs may help determine the right fit.

The main challenge isn't in choosing a club for ham radio, however, but choosing the one that is most appropriate. Except for clubs that emphasize individual involvement, like the public service clubs, you could join the number of clubs one wants to learn more about particular aspects of ham radio. Many clubs provide online publications and newsletters which provide information about the unique aspects of their clubs.

Following the joining of a general interest Ham radio group, active participation is encouraged via taking part in meetings, making new friends, and helping in the preparation of events and clean-up. The amount of benefits derived from the group is proportional with the degree of involvement.

How To Promote Your Channel?

Feedback from the Internet offers new possibilities Control of the site's content by users of the website, involvement in the creation of the content, posing challenges to coverage and debate and a proactive role in the debate of the source and its creator, the sharing of opinions on different issues with a journalist as well as others. The development of new communication methods is due to this method in addition to new ways of collecting data and analysing it are gaining traction.

The current situation is unstable within the FM spectrum of today and is constantly changing, which means the radio station has to find new methods to remain relevant in the current media marketplace. There is a need to find innovative methods to market your products, and due to this, there's a limited amount of time to air, which is why radio stations are able to continue to operate. They interact with their listeners in a variety of ways, including off radio. For a radio station to succeed today radio stations need to preserve its brand image and enhance the quality of its programming through three different ways:

1. In the role of a media outlet

2. It is a business venture that can be an effective marketing medium

3. In its capacity as an organization acting for the benefit of its viewers

For all of these aspects in all these areas, it's important to hear feedback from listeners, examine the market and know who the radio station is aiming at and the segment of that public is needed to attract to increase its growth. In this way we are able to pinpoint the most effective ways to use on-air communications with listeners as well as promotion for the radio station.

1. Advertising

2. Public relations

3. Promo campaigns

4. Public events that are held

5. Making use of Internet to market your business.

Selecting The Antenna For Radio Station

If you purchase or construct an antenna, you want that it will function efficiently and the signal it produces will surpass the other antennas. What is the ideal antenna to get what you want from it? The Nobel radio amateur is aware of the primary function of antennas is to transmit an electromagnetic signal that travels across the space. However, he doesn't understand how this feat of physics takes place. Before I get into the topic let me be clear that there aren't any magical antennas that are neither great, neither bad nor better and all of them do their job as they are given. When they fail to meet the standards, it might be due to an unwise choice, or incorrect calculation of mathematical formulas.

Radio stations no matter if it's amateur or commercial radio to make it efficient technicians for this instance that is your radio amateur, must be able to perform the necessary calculations for balancing the impedance the antenna produces and the correct elevation of the tower as well as the duration of the transmission cable (coaxial cable) in order to send to space the entire power that the transmitter produces and benefiting from the gains of the antenna.

Radio equipment is a part of " electronic devices, specifically the radio wave branch (wireless communications). In particular, the method for radio broadcasting the apparatus used to broadcast as well as receive radio signals. the radio transmitter, transmission line and antennas. However, antennas play a role in the magic of wireless communication through the emission and reception of waves. The dipole is certainly among the antennas which are at the top of the ladder (the class) of choice due to its performance/cost ratio. While it's not the only antenna amateur radios are planning to put in place to function in the HF band when they put the theories into action there are some intriguing details to warrant discussion.

To allow an amateur radio station become effective, its operator should apply the minimum information about radio technologies into practice

1. Conditions of operation (regarding dimensions.

2. Height above ground level.

3. Polarization

4. Standing Wave Ratio (SWR)

5. The type of soil

Separation from other antennas or elements that are constructive to the proper operation of the antennas. Antennas for HF bands. Based on the information provided in the background, we'll be referring to the HF antennas (3-30 MHz) that are those that need most space and the largest length. To begin one must point to the fact that the length of an antenna directly relates to the frequency (I) that corresponds to the band the frequency band it is required to fulfill its purpose. This makes it difficult to decrease the size of an antenna without sacrificing its effectiveness.

Take into account: height + altitude.

An antenna's height should be in line and operating frequency. The antenna will be more effective if it is higher in altitude above the ground as well as sea level, and also the more free it is from obstacles surrounding the antenna. The elevation above the sea level of an antenna is a factor that radio amateurs are not able to choose since every one of them must set it wherever they are. However they are able to alter the elevation of the ground and raise the antenna as high as they can over the surface. By using towers or masts. It is important to put the antenna as high feasible, particularly for frequencies between 14 and 15 MHz. It is recommended to place the antenna between half wavelengths and the full wavelength. For instance, if you are in the frequency of 14 MHz, or a 20 meter band, the optimal minimum height would be 10 meters, and the highest 20 meters.

Chapter 4: Exploring The World Of Amateur Radio

Ham radio is a fascinating environment where people exchange information using their expertise, creativity and enthusiasm. It has been in existence for over a century and lets people connect with each other across vast the distances. This article will discuss the development of amateur radio as well as its various potential. Ham radio was born in the latter part of 19th century when researchers and inventors such as Guglielmo Marcani began to experiment in wireless communications. The pioneers of the early days laid the basis radio technology, which would eventually result in amateur radio.

At the beginning of the 20th century the governments started to recognize the significance of radio communications to serve both civil and military goals. To make sure that radio spectrum was effective and well-organized usage of the radio spectrum the government started assigning frequencies and licensing the operators. Also, there were those who had an interest in radio communications and research, going far beyond government or commercial uses. They wanted to investigate radio technology on their own terms.

Amateur radio first came to the United States when a group of radio enthusiasts embraced this technology at an early stage. They invented their own radio equipment that shared data and used bidirectional communications in order to enjoy themselves.

The phrase "ham" originates in "ham-fisted," which was utilized to denote people with less skilled capabilities. In time however"ham" was adopted as a symbol for radio amateurs, indicating their enthusiasm for experimentation as well as dedication to their craft.

Amateur radio operators are one of the most important aspects of Ham radio. A person authorized by the authorities in their country to run radio equipment on the amateur band, plays an essential component of Ham radio. To be able to get a license you need to take an exam to test your knowledge of radio theories and rules.

Amateur radio operators enjoy specific rights. They can transmit on different frequencies, experiment with different methods of communication and participate in various events and contests. Radio operators who are amateur are able to connect to local network, as well as national and international networks, which allow they to connect with radio operators around the globe.

The capability of Ham radios to join people at long distances is just one of its most distinctive characteristics. The equipment used is usually modest which has very low energy. The ionosphere, as well as Earth's surface are able to bounce radio signals, which allows the signals to travel a lengthy distance. It allows for communication across countries and continents.

DXing is a method for creating long-distance communications that has caught the interest and imagination of generations of amateur radio users. It opens up a whole new realm of cooperation, friendships and exchange of culture that goes far beyond language and borders. Ham radio enthusiasts relish the excitement of connecting even when they are in far-off or remote areas. They accumulate QSL cards as proof of their achievements.

Ham radio is a varied activity that is able to be adapted to meet a wide range of abilities and interests. Many people build their own radios as well as antennas, and immersing their minds in electronics engineering and design. Certain operators concentrate on emergency communications, providing crucial communications connections during emergencies in which traditional infrastructure fails.

Radio amateurs also participate in events for the community, such as marathons, parades and gatherings with the community. In these events, they provide aid in communications as well as support. Also, there are contests as well as awards programs to encourage radio operators to meet specific objectives. In the case of example, they could be pushed to achieve the most contacts within an allotted time duration or reach any country around the globe.

There is a possibility that Ham radio still has a place in the modern world of hyperconnected technology with smartphones, internet, as well as other types of communications have taken over. Radio amateurs still play an essential role in our society. in times of crisis Ham radio operators provide an essential resource for those who require communication and relay critical data.

Ham radio can also be an excellent way to encourage technology-related education and to encourage innovation. Ham radio is an excellent opportunity to educate students to become engineers, technologists as well as scientists. They can be inspired to seek careers that are related to disciplines like radio communications. The capacity of radio communication to research and resolve problems could be extremely beneficial for society.

The amateur radio spectrum is barely scratched radio's vast tapestry of broadcasting as we end this chapter. Ham radio has been awe-inspiring to people over the years with its enthralling possibilities and the magic.

The details of the technical in the next chapters. From the understanding of radio waves to the operation of equipment and more, we'll take a look. Then we will look at the particulars of antenna design, Morse Code, as well as the digital realm. Also, we will discuss international associations that regulate amateur radio.

Get your seatbelt buckled and prepare to embark on an exciting journey into the realm of Ham Radio. We wish you the best, whether have experience or are novice to the hobby this website will pique you to explore and spark your interest in this unique method of communicating.

Chapter 5: Understanding Radio Waves: Principles And Propagation

It is essential to know about radio waves so that you can be able to appreciate the full spectrum of radio amateur. Radio waves are electromagnetic radiation which encompasses numerous wavelengths and frequencies. In this spectrum, you will find electromagnetic waves (radio), microwaves visible light, infrared light, UV X-rays, as well as Gamma Rays.

The electromagnetic waves comprise two components: electromagnetic and an electric field. Space is where they are able to oscillate perpendicularly. Because they don't require any physical media for transmission They can even travel through space that is empty, as in the outer space.

The frequency and wavelength of radio waves decide the way they operate. The frequency, expressed in Hertz is the amount of complete cycles in a second. The wavelength represents how far between consecutive peak and troughs. They are both inversely connected to shorter wavelengths, which are related to more frequency.

The Electromagnetic Spectrum

The electromagnetic spectrum covers many wavelengths, from radio waves to gamma-rays. Different parts of this spectrum possess different characteristics and functions. This is why it's crucial for operators of ham radio to be aware of these sections so that they can navigate and utilize the different frequencies of radio.

The radio amateur bands that are of interest span from just a few Kilohertz all the way to hundreds of Gigahertz. The bands can be classified into high frequency (high frequency), VHF(very high frequency) as well as UHF (ultrahigh frequency) as well as microwave bands. Each band comes with its own distinct set of characteristics in terms of transmission. It is the method by which radio waves traverse the air as well as interact with surroundings.

The frequency of the HF band is between 3 to 30 megahertz. It is the HF band is well-known due to its capability to travel far distances via reflection off of the Ionosphere. It allows radio amateurs across the world to connect. The propagation of radio signals can be affected by a variety of variables, including the time of the day, sunspots on Earth, as well as the magnetic fields.

The VHF spectrum is more prominent in the spectrum, and it covers the frequencies from 30 MHz up between 300 and 30 MHz. VHF signals are less length than HF signals and can consequently be transmitted via line-of-sight. VHF signals move along a straight line and only impacted by obstructions like structures and the terrain. VHF is utilized to transmit information in short distances, for instance, in regions.

The UHF spectrum spans from 300 MHz to 3 gigahertz. UHF signal strength is lower than VHF and mostly line-of sight. They are particularly useful for urban environments, since construction and structures can dramatically decrease the strength of signals at low frequency. UHF frequencies can be used in applications like two-way radios as well as mobile communications devices.

The spectrum of microwaves extends beyond UHF, and encompass frequency ranges between 3 GHz up to a few gigahertz. The microwave signals have a very short wavelength and are primarily used for the transmission of high-speed data satellite communications, as well as radar technology. They are extremely directed and need precise alignment of transmitting antennas and the receiver antennas.

Ham radio operators are able to select the most suitable radio frequency band by analyzing the characteristics and capabilities of every band. This allows them to pick the frequency best suited to the particular application or environment they're seeking. Ham radio operators that adapt to the specific characteristics of every band will increase the efficiency of their communications. and reliable communication, no matter if they're for conversations in the local area as well as long-distance contact or for specialized microwave communications.

Radio Wave Propagation

Radio wave propagation refers to the process by the radio waves interact with their surroundings and spread out into space. The reliability, quality, and the range of communications are dependent on the propagation process. Understanding the different mechanisms of propagation helps radio operators to optimize their processes and assure efficient communication. Radio waves are influenced by a variety of key processes. The mechanisms are different based on the frequency band, as well as specifics of the area in which radio waves function. In this article, we will look at some of the most important transmission mechanisms.

Ground Wave Propagation

Groundwave propagation is an essential process by that radio waves travel and interact with their surroundings. It is primarily observed in lower frequencies, usually with several megahertz (MHz). Understanding how groundwave propagation works is vital in optimizing radio communications particularly for transmissions with short range. In this article, we'll explore the fundamental concepts and features of groundwave propagation.

When radio waves reach us via the antenna they are radiated across the entire world. Once they reach the surface of Earth and interact with it, they and follow its curvature radiating along the earth. The interaction between radio waves and surface of the Earth is the foundation of groundwave propagation.

One of the main aspects that affect groundwave propagation the frequency of radio waves. In lower frequencies, like those in the AM broadcast spectrum (around 800 kHz to 540 kmhz) the propagation of groundwaves is the primary method. This is because of the extremely long wavelength of these frequencies. This permits radio waves to be more effective in their interactions with earth's surface.

The terrain's characteristics and the conductivity of ground can also affect the groundwave's propagation. The surface of the Earth can be considered as a multifaceted medium composed of various kinds of terrains, such as water, land, and vegetation. Each type of terrain have different properties of conductivity that affect the behaviour of radio waves.

As radio waves pass through the surface of the Earth, various factors take place that influence the propagation of groundwaves. This includes diffraction, reflection, as well as absorption. Let's look into each in greater detail:

Diffraction: Diffraction is the bend of radio waves over obstacles that block their way. When radio waves encounter obstacles, like the top of a building or hill the wave diffracts or curves around it and spreads into different directions. The bend allows radio waves to travel into regions that aren't in the direct line of sight of the transmitter antenna. As a result, groundwave propagation enables communication in non-line-of-sight scenarios.

The degree of diffraction is contingent on a number of variables, including the frequency of the radio waves and the shape and size of the obstruction. In general, wavelengths with longer wavelengths scatter greater than shorter wavelengths. That's why the propagation of groundwaves occurs more strongly in lower frequencies, where wavelengths are greater.

Reflection is a phenomenon that occurs when radio waves hit an electrically conducting surface for example, the Earth's surfaces or artificial objects. The incident reflection is returned to space, and the remainder is directed towards the ground or is absorbed. The reflection effect has a major impact on the propagation of groundwaves, since it permits radio waves to travel across greater distances.

The reflection of radio waves can be dependent on a variety of factors that include the angle at which they strike radio waves, the surface which reflects them and the frequency they are reflected off. Since the Earth's surface is an excellent conductor for radio waves, they reflect much of it of them back to space. The direction in which the radio waves are reflectors is determined by the direction at which radio waves touch the surface of the Earth.

Absorption is the process by which radio waves absorb the surface of Earth or by other elements within the propagation path. If radio waves come into contact with the earth or with other substances the energy transforms into heat. The absorption rate of radio waves is dependent on the dielectric property and conductivity of the substances they come into contact with.

Different kinds of terrains and materials exhibit different degrees of absorption. Dry soil, for instance, is more absorbent in comparison to moist soils or water bodies. Plants, buildings and various other structures may absorb radio waves to various extents, according to their size and composition. Absorption decreases the power of the radio waves as well as reduces the distance of propagation by groundwaves.

Additionally to these issues in addition, the effectiveness of propagation by ground waves can be affected by the atmospheric conditions. The variations in conductivity as well as temperature and humidity may cause changes in the characteristics of propagation for radio waves.

The range of groundwaves is contingent upon a variety of variables such as the frequency and power of radio waves, as and the conductivity. When frequency rises, the groundwave propagation is reduced as other methods of propagation including skywave propagation, are more dominant.

For optimal groundwave propagation to ensure optimal communication, radio engineers are able to employ a range of strategies. This includes selecting the right frequency as well as optimizing the antenna's size and location, as well as looking at the terrain and ground properties at both the receiving and transmitting ends. Modern methods of modeling and measuring are available to help predict and evaluate the groundwave propagation properties in specific areas.

The groundwave propagation is the primary mechanism used for radio communication in the short range, especially in lower frequency. It is the result of interactions between radio waves and the earth's surface. This causes effects like diffraction, reflection and absorption. Recognizing the nature and characteristics of propagation by groundwaves helps radio operators optimize their operation and guarantee high-quality communication in short distances.

Skywave Propagation

Skywave propagation plays an important function in communications over long distances using the HF band. Radio waves bend or are focused by the ionosphere which is a part of the upper air of Earth which is home to ionized particles. Ionospheres have several layers that include the F1 layer as well as F2 layer. F2 layer.

It absorbs radio waves that are of less frequency during the day and blocks skywave propagation. The D layer disappears after sunset. This enables the E and F layers to transmit HF signals back to Earth. This allows long-distance communications across thousands of kilometers and occasionally even continents.

Tropospheric Propagation

The tropospheric spread is known as tropospheric conduiting. It takes place within the lower layer of Earth in the troposphere. Refraction and bends that result from fluctuations in temperature and humidity are what make it unique.

Tropospheric ducting can be particularly important particularly in those VHF (Very High Frequency) and UHF (Ultra High Frequency) bands, in which it is able to enhance the signals and allow long-distance communications beyond normal lines-of-sight restrictions. The mechanism can be seen in the case of temperature inversions where warm air is encased by the cooler air.

In the course of a temperature change that radio waves are exposed to, they experience an alteration in refractive index the atmospheric layer. This causes them to bend in the direction of curve of the temperature-inversion layer. This allows the waves to travel for longer distances. The effect of ducting allows radio waves to extend beyond what is normally achievable using the principle of line-of-sight.

The tropospheric ducting phenomenon is most commonly seen over bodies of water, including oceans or huge lakes, in which significant temperatures can be observed.

The difference between the warm air above the lake and the cooler air above the ground creates a perfect conditions for tropospheric conduction to take place.

In these regions radio waves may travel along the curvature in the inversion layer that controls temperature. They move in a similar way to ducts expanding their reach and making it possible to communicate with distant places which would otherwise be blocked by Earth's curvature, and other obstacles.

Tropospheric ducting has both positive and negative consequences on radio communications. One way is that it could boost signal strength, and facilitate communication over long distances that otherwise would not be possible or difficult. But there are also changes in signal strength and quality as conductor's conditions change quickly because of atmospheric conditions.

The tropospheric process also known as tropospheric ducting is a radiowave propagation technique that is found in the lower regions of Earth's atmosphere. It is the process of bending or reflection of radio waves as a result of changes in humidity and temperature. Tropospheric conduits can increase the reach of VHF and UHF communication, making radio waves move in a similar way to ducts and especially in locations where there are temperature inversions or substantial temperature variations. Utilizing tropospheric conduits could be beneficial to long-distance communications in specific frequencies and geographic locations.

Sporadic E Spread

The phenomenon of sporadic E (Es) spread is an intriguing phenomena that occurs within the E layer of Earth's Ionosphere. It happens when areas of ionization, also known as the sporadic E clouds, develop in the air. These clouds serve as reflective surface that transmit VHF (Very High Frequency) as well as UHF (Ultra High Frequency) signals, which allows the signals to travel across medium-long distances.

One of the most fascinating characteristics of this type of E propagation is the impact it has on communications. As these scattered E clouds develop it creates opportunities for better signal reflection. This allows radio operators to establish connections that are typically beyond their normal reach. This could be especially useful for amateur radio users who love exploring the possibilities of long-distance communications.

In contrast, the sporadic E propagation can be less frequent and more localized in the summertime. That means during this time radio stations could encounter a greater frequency of scattered E cloud formations. This leads to a greater number of chances for communication over long distances. Nature is like giving the radio listeners throughout the summer months.

In short in summary, the sporadic E propagation happens in the form of fragments of ionization that are irregular that are referred to as the sporadic E clouds, develop within the E layer of the Ionosphere. These clouds serve as reflectors of VHF as well as UHF signal, which allows to improve communication across medium-to-long distances. Particularly in summer the sporadic E propagation can be more common and limited giving amateur radio operators an opportunity to make contacts outside of their typical distance.

Meteor Scatter Propagation

Meteor scatter propagation utilizes the ionization trails left by meteors after they have entered the Earth's atmosphere. When a meteor is in the atmosphere of Earth it is ionized by air molecules throughout the space, resulting in an ionized trail. The ionized path can be utilized to reflect VHF as well as other UHF signals. It permits communication between many hundreds of kilometers.

Ham radio operators can make use of meteor showers as well as random meteors to bounce signals from the ionsized trails. It will increase their communication reach. To decode and detect the weak signals that reflect off meteor trails, special tools and software are typically employed.

Line-of-Sight Propagation

Line-of-sight is the most efficient method for transmitting radio waves. Radio waves move along a straight line from the antennas which transmit to receivers without reflections or obstructions. Line-of-sight communications are limited due to the curvature and shape of the Earth obstructions like mountains, buildings and plants also.

Line-of-sight communication is subject to a range of variables, such as the height of antennas, their the power of transmission, gain frequency, as well as the geography. Ham radio operators employ repeaters to extend the reach of their line-of sight communications. Repeaters are stations strategically placed which receive signals and then retransmit signals to increase their range.

The factors that affect the Radio Wave Propagation

Radio wave behavior are influenced by a range of elements. Ham radio users can enhance their communication by analyzing these aspects and adapting their communications to changes in the environment.

Frequency

The frequency of radio waves determines the way it propagates and how it interacts with the environment. Ham radio operators should be aware of the characteristics of these waves for planning and optimizing communication.

Higher frequencies are associated with low-frequency radio waves for instance, those that are found within the MF (medium frequency) as well as LF bands. Longer wavelengths allow the radio waves to travel farther through various methods.

Ground waves are a key process in the transmission of signals at low frequencies. Ground waves are a result of the curvature and contour of the earth's surface. They are able to travel for across long distances. Typically, they travel as long as several hundred kilometers. The earth's conductivity and its frequency are the main elements that affect the mechanism of propagation. The lower frequency waves penetrate the earth faster and are able to get over obstacles and are able to reach receivers farther away.

Another technique for low-frequency signals be transmitted is using skywaves. Skywave propagation takes place in the band of high frequencies (HF) which occurs when radio waves get bent due to Ionosphere. It bends radio signals so that they can be able to reach Earth's surface from the distance of. It is better at being interacting with low frequencies of the frequency band HF, like the ones used in amateur radio. This leads to greater effectiveness of propagation of the skywave. It allows for long-distance communications over hundreds of miles or continents.

The wavelengths of high-frequency radio waves, for instance ones found in VHF UHF and microwave band, are smaller. These shorter wavelengths of radio waves makes them more vulnerable to effects from the atmosphere and obstacles.

When frequencies increase, the line-of-sight increases in importance. Line-of sight is the term used to describe radio waves that are in a straight line between antennas for receiving and transmitting. The curvature of the Earth, buildings, hills and the vegetation form the primary barriers. These obstructions are more likely to affect or weaken high-frequency signals. Line-of sight communications typically limit by a few tens of kilometers.

When frequencies are higher, atmospheric influences like absorption, reflection and scattering become significantly more pronounced. This can cause more attenuation in signals, and a decrease in. For instance rain or fog, they can absorption and scattering of high-frequency signals which can result in decreased signal clarity and the strength.

Ham radio operators utilize methods like antenna high optimization, signal amplifying as well as directional antennas in order to get around the limitation of high-frequency propagation. They can guarantee reliable communications by selecting the correct frequency band, based on their desired range and the surrounding circumstances.

The frequency of the wave could have an effect on the characteristics of propagation. Radio waves with lower frequencies can make use of skywave propagation for communication across vast distances. Signals with higher frequencies are less susceptible to the effects of atmospheric or obstructions and atmospheric effects, while lower-frequency signals tend to be more vulnerable. When selecting a frequency band for Ham radio, operators must consider these aspects.

Size and Gain of the Antenna The fact is that height and gain play a major role on the transmission of radio waves. Ham radio operators who take into account these aspects can improve their communications by strategically altering their antenna's height, and choosing antennas that offer more gains.

The antenna's height can have a major impact on the frequency range of radio waves that are able to be observed. This is especially relevant for frequencies within the VHF (very-high-frequency), UHF, and microwave bands where line-of-sight propagation is dominant. Radio waves travel much more quickly if an antenna is elevated. They can thus avoid obstructions such as trees, buildings as well as the terrain. The range of communications between transmitters and receiving stations increases because of this free pathway. The curvature of the Earth is less of a barrier and the more powerful the antenna and the greater the line-of-sight distance. Affixing antennas to high structures or on elevated structures could increase the reach and range of communication.

An antenna's gain is equally important as elevation of radio wave propagation. Antenna Gain refers to the antenna's capacity to concentrate or direct the energy radiated to a specific direction. An antenna's gain can be measured in decibels. It can be used to determine the directivity of an antenna. The antennas with higher gain are able to transmit and receive signals over greater distances, and with greater the strength of signals.

High-gain antennas boost the directivity of their beam patterns. This helps to concentrate the energy into an emitted beam. Concentration increases, allowing the signal to be directed to the receiver you prefer which increases reception distance and quality. High-gain antennas may be utilized in applications that require long distance communication or signals that need to be aimed in a certain direction, such as point-to-point communications or satellite communications.

It is important to note that high-gain antennas have a cost and offer greater coverage, but they come at the expense of different direction. They provide greater coverage as the pattern of radiation narrows. In selecting and locating the high-gain antennas it is essential to think about the specifications of the particular communication situation.

Ham radio enthusiasts can boost their communication and signal strength distance by carefully adjusting their the height of their antennas. The combination of higher heights to increase lines-of-sight with high-gain to improve transmissions and receptions provides greater coverage, more communication distances and overall efficiency. These aspects are particularly relevant in situations where reliable communications are required across long distances, challenging environments, or to support satellite communication.

Atmospheric Conditions

Relative humidity and temperature are the two air conditions that could affect the propagation of radio waves. Radio waves are affected by refractive indexes that are present within the air.

Radio waves' paths is affected through temperature changes in air. Inversions in temperature are an obvious effect in which hot air rises over cool air close to the surface of Earth. Temperature shifts that are inverted can result in the phenomenon of tropospheric dilation. In this case, radio waves bend or are focused along an inversion. This allows radio waves to extend far beyond the normal line of view, which allows communication over long distances.

Another parameter in the atmosphere, humidity can affect the propagation of radio waves. The dielectric properties in the atmosphere may be affected by variations in humidity. This in turn influences the rate and frequency of radio signals. Higher humidity levels will result in greater distorting and attenuation of signals which results in weaker and less effective signal strength. Airborne water vapor will reduce the intensity and clearness of radio waves in higher frequency.

The same is true for changes in the atmospheric conditions could influence radio waves' propagation. The air's density could be affected by variations in atmospheric pressure. This could affect the speed as well as frequency of radio waves. Pressure variations can cause slight deviations in transmission of radio signals.

Ham radio operators need to be in a position to detect and comprehend these conditions. Weather reports, models of the atmosphere as well as real-time data are a good way to evaluate the weather conditions and allow operators to take informed choices about their communications strategy. The operator can improve their communications performance by altering the strength of their transmissions or frequency, as well as antenna settings based on the humidity and temperature levels.

Abstract: The atmospheric conditions, like humidity and temperature can significantly affect the propagation of radio waves. Inversions in temperature can affect the direction of radio waves, which can allow the transmission of long distance communications. Signal distortion, attenuation as well as overall strength are affected by variations in the atmospheric or humidity conditions. Ham radio operators can adjust their radio systems according to changing atmospheric conditions through monitoring and analyzing weather conditions.

Solar Activity

The sunspot cycle can have significant effects on the propagation of skywaves across the high frequency band. The amount and intensity of sunspots impact the degree of ionization within the Ionosphere. It directly affects the capability radio waves to reflect and reflected back towards Earth. Skywave propagation is increased during times that have high solar activity and allows for communications over long distances. Skywave propagation is limited when there is low solar activity.

Ham radio operators need to understand the fundamentals and processes that drive radio waves. The knowledge they gain allows them to pick the most appropriate frequency and to adapt to changes in the environment to make effective communications. Every propagation technique offers amateur radio users an opportunity to study radio-based communication. From tropospheric e and intermittent to ground wave and skywave, provide unique experiences.

Chapter 6: Navigating The Technician Class Exam

The Technician Class license is the ideal option to start your journey in the field of amateur radio. This Technician Class license permits users to operate on a particular spectrum of frequencies as well as connect with fellow amateur radio operators. In order to obtain the Technician Class License, a person has to pass a license test that test their understanding of radio concepts, laws and operating protocols. The basics of electronics also are tested. The chapter contains valuable learning strategies as well as vital tools to aid aspiring users get through an exam called the Technician class Exam.

Understanding Exam Structure

The Technician Class Exam is comprised of questions that offer many options covering various subjects regarding amateur radio. The questions aim to assess the ability of candidates to comprehend fundamental concepts and how they can be applied in real-world scenarios.

The quantity of questions along with the marks needed for passing the exam will differ based upon the location where the exam takes place. Contact the appropriate regulatory agency or the amateur radio organization to determine the precise specifications and guidelines.

Study Tips for the Technician Exam

This Technician Class Exam is a test that requires dedication, focus as well as a systematic method. These strategies for study can help you to maximize the amount of knowledge you acquire to improve your chance of passing.

Be familiar With Exam topics. Know the outline of the exam as well as the subjects that are covered. The topics include radio theory, security, antenna fundamentals electronic basics, fundamental electronics, rules and operating protocols. Create a complete study plan covering all the topics.

Use Study Guides. Study guides that are specifically tailored to prepare you for those taking the Technician Class Exam can be a valuable resource. They provide a systematic approach to studying, and also include several practice questions, as well as detailed explanations. They can assist you to establish a strong foundation of knowledge and strengthen the fundamental notions.

You can join Study Groups Study partners: Finding a study companion or enrolling in a class will enhance your experience of learning. Being with other students gives you an opportunity to understand concepts by sharing resources and also ask questions. The ability to share what you've learned helps to strengthen your knowledge.

Practice by taking sample tests The sample tests are an excellent method to get familiar with the structure and contents of the exam. This can allow you to determine areas that require more study and will also help you build confidence. A lot of websites and amateur radio organizations offer exams for practice.

Use Online Resources. The Internet provides a wealth of information available to amateur radio. Forums, websites as well as online communities which are devoted to amateur radio provide a variety of study materials, information as well as practice tests. Make use of these sources to improve the quality of your education.

Experience in the hands is crucial. Although theoretical understanding and hands-on experiences are equally important and can help you improve your knowledge of radio amateurs. Take part in radio-related activities like creating antennas or operating radios. Participating in amateur radio-related events, or joining clubs of amateur radio in your area could give you the chance to network and learn.

Recommended Resources for the Technician Exam

Take a look at the following resources to prepare you to take your Technician Exam: Class Exam:

1. ARRL Manual ARRL Manual American Radio Relay League publishes manuals specifically designed for every class of amateur license. This manual, titled Technician Class License Manual covers exam subjects in detail and provides practice tests with detailed clarifications. This manual is a fantastic study guide for getting ready for the exam.

2. Online study guides There are numerous sites that provide studies guides specifically designed to prepare you for those taking the Technician class Exam. The guides provide thorough details of the various topics on the exam, test questions, as well as tips for success. They can prove to add value to the study plan.

3. Mobile apps provide flashcards, study materials as well as practice tests to help you pass your Technician Class Ham Radio License. They allow you to study and test your understanding while on the go.

4. Online Tests for Practice: A variety of websites provide online test practice exams which mimic the structure and content of the real exam. The practice tests will allow you determine the level of your preparation and identify areas that require improvements, and increase confidence.

5. Amateur Radio Forums & Communities Participating in the amateur radio community online is a fantastic opportunity to receive help and advice. There is the possibility of interacting with other radio operators by taking part in forums online and community.

Exam Day Tips

It's important to stay at ease and well-prepared before the exam. These are some helpful tips to assist you in doing the best you can.

1. You should rest well in the evening before taking your exam. Minds that are well rested will be more focus and sharp during tests.

2. Be early to arrive: Give you plenty of time to complete all administration or paperwork. Stress can lead to anxiety, which can negatively impact the quality of your work.

3. Make Time to Learn the Answers Take your time to take your time reading each question and attentively. It will assist you in comprehend what's being requested. Be aware of specific words and terms that can impact the answers.

4. Do what you are familiar with If the question is confusing or complicated, you can proceed to the next one and come back to it afterward. For greater confidence and

momentum you can answer any questions you're confident with.

5. Be organized Control your Time: Be sure to pace yourself throughout the exam to ensure you have enough time to complete the whole list of questions. Do not spend too much time on one question that is difficult, because it could decrease the time that you are given for the other questions.

6. It's been a long and hard process to get ready in this exam. Keep calm, trust your abilities, and be able to answer every question with confidence.

The Technician Class exam is an excellent opportunity to begin your journey with amateur radio. It is possible to pass the Technician Class exam with the help of instructions for studying and utilizing suggested sources. The license allows for communications, exploration and collaboration within the hobby of amateur radio.

Chapter 7: Operating Procedures And Etiquette

Etiquette and operating protocols are vital elements of radio amateur. Following the correct procedures will ensure the most efficient and pleasant communication on the radio waves. This chapter focuses on the essential operating procedures and manners of conduct that each amateur radio operator needs to know.

Basic Operating Procedures

Call Signs: in the world of radio amateur communication calls are very vital. They're unique identifiers that are assigned to radio amateurs that enable them to be identifiable and connect with specific operators. The most common method to begin conversations is by using your name, followed by the name of the station the station you're attempting to join.

Imagine yourself in a crowded area, attempting to catch the attention of someone. Instead of shouting at everybody, you can make use of their name and address directly. Call signposts, too, can be used to identify and distinguish specific operators within the large community of amateur radio operators.

When you initiate the transmission, you typically announce your name first to inform others of the person you're. This allows other operators to recognize your station and respond with the appropriate response. After your call signal then you should use the call sign for the station you're trying to reach, indicating that you are willing to talk with them.

It could be something like, "KD2XYZ, this is W1ABC." When you announce your call number before you say anything, you are telling an operator using the sign KD2XYZ that you're speaking to them directly. The caller with the sign KD2XYZ is then able to listen attentively when you spoke to them.

The incorporation of call sign language into radio messages aids the development of a clear and organized system for creating and keeping contacts. It ensures the messages reach the correct recipients, and helps reduce confusion for the operator.

Call signs are unique identification codes which amateur radio operators utilize. The past was when it was standard to initiate conversations by saying your call number, and later the name of the station you attempted to connect with. This helps in the finding of particular radio operators, and also creates a clear communication line within the radio amateur community.

When you are listening before making any transmissions via radio you must practice appropriate radio manners by watching. Making sure you are listening to the frequency that you would like to use, and then carefully watching the conversations in progress are necessary. This will ensure that the frequency you want to use isn't being used, and hinders interference from the ongoing conversation.

The practice of listening before you send has several purpose. First it helps you find out if the frequency is available for usage, and ensures you don't accidentally disrupt the ongoing conversations or interrupt others' communication. It also encourages radio users to cooperate and respect each other.

In addition, by listening to the sound and frequency, you will become more familiar with ongoing conversations and conversations. This helps you understand the meaning and context of discussions that are ongoing and makes it simpler to be a part of the discussion or reply appropriately when sending your personal messages.

To avoid interrupting current exchanges be sure to listen for a suitable period of time prior to transmitting. It shows respect and tolerance to other users' ongoing conversation. Through attentive listening and not interrupting someone else's conversation or interrupting conversation.

In summary, it's vital to engage in active listening prior to sending any messages. This involves tuning into the frequency and keeping tabs on current discussions. It helps to confirm that the frequency is in use and also to be aware of recent transactions. If you are listening for the appropriate duration of time it shows respect for ongoing communications, and reduce the need for interruptions. Listening skills that are good are crucial to radio etiquette since they allow for effective and harmonious communication between radio broadcasters.

Q-Signals: Q signals are a standard set of three-letter codes utilized for amateur radio communication to provide specific information in a short and concise way. They can streamline communications and help save time. Learn about commonly-used

Q-signals, and the meanings they convey. Examples: "QSL" means "I acknowledge receipt" or "message received." "QRM" refers to an interference. "QRZ," asks, "Who is calling me?"

PTT (Push-to-Talk) Once you're prepared to transmit your message, you must that you follow the proper procedures to ensure that you're communicating effectively. A crucial process is to turn on your transmitter using the push-to talk (PTT) button in your transceiver or microphone.

The PTT button acts as a radio's device as a control mechanism. It is activated when you press and hold the PTT button. This allows the message you speak to be sent to the others using the identical frequency. The message will be heard clearly by those who you would like to connect.

But, remember that the release of the PTT button needs to be handled with care. Before you press the button, be sure you've completed your message and include all essential details. It includes your call number and the call number of the station that you're calling along with any other pertinent information or information.

The release of the PTT button too quickly could result in a cut-off or incomplete messages, which can make other operators confused if they are watching. Make sure you give a clear and complete message by making sure that you've completed your message before you release the PTT button. This allows the other operator to read it and act in a timely manner.

It's important to remember that using the PTT button and the appropriate microphone method can aid in communicating more efficiently. It involves speaking clearly at the right volume. It also involves making sure that you are free of background noise and interruptions that could affect the clarity of your communication.

In short, it's crucial to press the push-to talk (PTT) button in your transceiver or microphone to activate the transmitter while you are sending a message. Pressing the PTT button will allow the message to be transmitted to the world at large. But, it's important to let go of the PTT button only after having completed your communications, including the call number and any other details that might be needed. Your message will be comprehended by the other operator and allows for efficient communications over radio waves.

Full Break and Part Break If you are joining in on a conversation, you must to declare your presence. Make use of "full break" or "partial break" to signal your intent to add your voice. Full breaks mean that you are making a big announcement or an emergency. On the other hand, the term "partial break" indicates an interruption of a short duration. Example "Full break. This is KD2XYZ and an emergency. Over."

Net Operation: Nets play a significant role within the field of amateur radio as they act as an organized gathering for radio operators who have distinct objectives. They can be aimed at communicating information and discussing specific topics to providing an area for communication in emergency situations. In order to facilitate smooth and efficient communications, participants in nets must adhere to specific guidelines.

It is first and foremost that it is crucial to adhere to the instructions of the net controller. Net controllers are responsible for making sure the network is in good order as well as managing the flow of information. They could set up specific protocols, establish time times for the members to exchange messages, or provide guidance on the general organization of the network.

It is essential to be patient and be patient and wait for your turn during the net. The web controller will alert you when it's safe to communicate. Attention and patience are crucial when you are listening to other people and wait until your turn.

The act of waiting for your turn to speak is a sign of respect for the internet's structure and guarantees that every person gets the same chance to hear. This helps to ensure a well-organized and efficient flow of data and maximizes the effectiveness of the internet to achieve its objectives.

Also, it's important to prepare before your turn to talk. Take your time and reflect about the message or facts you'd like to convey. Making sure you are concise and clear in your communications facilitates effective communication of information as well as aids in the smooth functioning that is the Internet.

The net controller's instructions gets more important when it comes to the case of emergency communication nets. A clear and organized communication system is crucial in times of crisis in order to coordinate the actions of others and offer help. The net controller's instruction and waiting to speak will ensure the emergency communication is handled correctly and effectively.

Nets, in essence, are groupings of amateur radio operators that have specific objectives like information sharing, or emergency communications. You should follow any instructions given by the net controller who is responsible for the flow of communications during nets. When you wait for your turn, it indicates respect for the organization of the net and helps facilitate a smooth and organized conversation. When you adhere to these rules will help the web to achieve its objectives more efficiently.

Emergency Communications: During emergency emergencies, appropriate operating protocols are vital. Utilize clear and succinct terminology, provide essential details, and obey the instructions of emergency coordinators. Be sure to keep the frequency free to avoid emergency communication and messages that are not essential. Operating Etiquette

Courtesy of Amateur radio is an sport that encourages the bond of camaraderie, as well as the respect of others. Be courteous to fellow radio operators and respect, no matter their level of experience or geographical area. Avoid taking part in heated debates or using language that is offensive.

Concise and Clear Communication Keep your communication clear when you communicate by speaking slowly, and clearly articulating your messages. Make use of plain language, and stay clear of technical terminology or jargon which may not be widely recognized. Make sure your message is clear in its delivery, delivering the information you need without unnecessary detail.

Listen first: Active listening is an essential element of proper operating. Prior to transmitting, you should ensure that you take time to take the time to observe and learn from the current conversations. Beware of interrupting conversations or creating excessive disturbance. Participate in discussions when it is appropriate and participate in meaningful ways.

Interference: Limit interference by using the correct frequency and by ensuring that your equipment works properly. If you accidentally interfere with the ongoing conversation, be sorry and revert to the frequency in order so that the conversation can carry on.

Elmering is a reference to the mentorship of experienced operators and helping new or less skilled operators. If you've learned something you wish to share, then be prepared to aid others in providing assistance and responding to questions. Encourage a friendly and warm environment within the radio amateur community.

Operating Band Plans: Operating bands provide guidelines regarding frequency usage in particular bands. Be familiar with the bands that are that apply to your particular license class and the region in which you reside. Follow the frequency and mode to ensure effective and efficient communications.

DX Code of Conduct DX (long-distance) communications pose particular problems. DX Code of Conduct: The DX Code of Conduct sets out the guidelines that apply to working DX stations that emphasize the importance of fairness, patience and the respect of others. Learn the DX Code of Conduct and follow its guidelines when participating with DX operation.

Emergency Communications and Public Service Events

Radio operators who are amateurs play a vital role in emergency and public service communications. In the event of participating in these events be sure to follow certain operating guidelines and rules of conduct:

Communication during emergencies: Efficacious communication is crucial to allow for an efficient response and coordination during emergencies. When faced with such situations it's crucial that radio enthusiasts using ham radios prioritize the communication of crucial data. Also, they must follow the particular protocol for communication established by emergency coordinators as well as the relevant authorities. They will help in ensuring that there is a smooth and well-organized emergency response, by following specific protocols of communication set by emergency coordinators and relevant authorities.

In the case of emergency communications in emergency situations, it's important to ensure that the message is simple and succinct. It is essential to convey details concisely and clearly to prevent confusion, and encourage rapid decision making. In times of emergency every minute counts. Operators can help emergency response coordinators and other responders to make informed choices by providing clear and concise update.

It is vital for ham radio operators to follow the rules and guidelines provided by the incident command or emergency coordinators during emergency situations. They possess the skills of and the experience required to oversee and coordinate responses. Following their guidelines they can guarantee an efficient response, and decrease the chance of miscommunications and contradicting details.

It is essential that users avoid sending information that is not essential during times of crisis. This helps to keep areas free of critical data and prioritised traffic. Conversations between friends, discussions that are not related as well as unnecessary chatter may be deemed to be not essential. The messages could disrupt the emergency operation or cause communications channels to get blocked. Ham radio operators who practice control can improve the effectiveness and reliability of emergency communications by eliminating excessive messages.

It's also crucial that organizations prioritize the sending and reception of important details. It is crucial to communicate important updates, safety guidelines emergencies, safety instructions, and any other information that is essential. The process of prioritizing information helps to make sure emergency workers will be able to take appropriate actions and effectively allocate resources.

Ham radio operators also communicate messages in times of crisis between organizations in response to the emergency and coordination groups. It is crucial to relay messages in a timely manner and speedily in these circumstances with no errors or mistakes. Operators should be active listeners and ensure that they have understood the meaning of the message. They also need to relay the message in a precise manner.

In the end, Ham radio operators are crucial for maintaining efficient communications in crises. They will aid in ensuring that the effectiveness of an disaster response by prioritizing critical information, and by ensuring clear and short communications. They are able to ensure it is synchronized and timely effective resource allocation as well as better results in challenging circumstances by adhering to efficient emergency communication.

Event of Public Service: Accurate communication is vital to ensure the safe and efficient running of public service activities like marathons or parades. Radio amateurs are vital to meet this need. They make use of their talents and expertise to establish solid communication channels for all of the occasion.

It is essential to adhere to the procedures offered by event organizers in the event of being a radio amateur at public-service events. These guidelines were created to facilitate efficiency in communication and coordination between participants, event personnel, as well as other important staff members. It is possible to contribute towards the overall success your event by adhering to these rules.

The ability to be a competent, professional radio operator in public events is vital. It is essential to keep an appropriate and courteous manner when dealing with organizers of the event, operators and attendees. Timeliness, professionalism, and planning demonstrate your dedication to the role you play. The radio you represent is amateur and are required to maintain a high standard of conduct throughout the day.

The reliability, in turn, is an essential characteristic of amateur radio operators engaged in public affairs. Your skills in communication and expertise can be relied on by the event organizers to make sure that the event is run efficiently. There may be a need to communicate important details, coordinate resources or provide information on the status of the occasion. The best way to boost confidence among the organizers by being trustworthy and dependable. This can also improve the security and effectiveness of the event.

www.ingramcontent.com/pod-product-compliance
Lightning Source LLC
Chambersburg PA
CBHW071334120626
46546CB00002B/558